Sublime Communication Technologies

Sublime Communication Technologies

Rod Giblett

palgrave
macmillan

First published 2008 by
PALGRAVE MACMILLAN
Houndmills, Basingstoke, Hampshire RG21 6XS and
175 Fifth Avenue, New York, N.Y. 10010
Companies and representatives throughout the world

PALGRAVE MACMILLAN is the global academic imprint of the Palgrave
Macmillan division of St. Martin's Press, LLC and of Palgrave Macmillan Ltd.
Macmillan® is a registered trademark in the United States, United Kingdom
and other countries. Palgrave is a registered trademark in the European
Union and other countries.

ISBN-13: 978–0–230–53743–9

This book is printed on paper suitable for recycling and made from fully
managed and sustained forest sources. Logging, pulping and manufacturing
processes are expected to conform to the environmental regulations of the
country of origin.

A catalogue record for this book is available from the British Library.

A catalog record for this book is available from the Library of Congress.

10 9 8 7 6 5 4 3 2 1
17 16 15 14 13 12 11 10 09 08

Transferred to Digital Printing 2012

Dedicated to Zoë Sofoulis

Contents

Preface

This book is a critical cultural history of communication technologies. The history it traces is a genealogy that follows the line of descent from railways and telegraphy to computers and the Internet. The central theme, and main object of critical analysis, is the sublime. For Benjamin (1999a, p. 415), the sublime is simply, and literally, 'carrying aloft'. Communication technologies quite literally carry information above the earth. Beginning with the railway and the telegraph, communication technologies carry information above the earth and through terrestrial space. Later with satellites they have carried it even further above the earth into extraterrestrial space.

Yet the sublime does not merely involve upward spatial displacement. It also carries a lot of metaphorical and ideological baggage. This includes the chemical process of sublimation of solid matter into gas. This process has also been used as a metaphor for mind–body dualism and the privileging of mind over body; for the transcendence of the immanent; and for the displacement of instinct into intellection and of place into space – all of which characterise communication technologies, including their invention, production and utilisation. Communication technologies attempted to transcend earthly life into a secular heaven devoid of God. The sublime is a secular theology, a modern mythology of technology.

Drawing on chemistry, Marx and Engels in the *Communist Manifesto* used the sublime without naming it as such to argue that in industrial capitalist societies 'all that is solid melts into air'. For them, the solidities of pre-capitalist social formations under industrial capitalism are sublimated into thin air and have an ethereal existence. Following a similar line, Berman in *All That is Solid Melts into Air*, took up Marx and Engels' phrase and explored the etherealising project of modernity. Yet he only discusses mass communication in general terms (1983, p. 16) and very briefly. He never considers the technologies of communication, a curious omission from a book devoted to discussing modernity in which those technologies have played such a crucial and formative part.

Communication technologies played a crucial role, not only in modernity generally as Thompson (1995) argues beginning with print, but also particularly in transforming the solidities of pre-capitalist social formations into thin air. This applies to the trajectory of development beginning with telegraphy and radio, through television and satellites and culminating at present in the Internet and the use of electromagnetic spectrum. Communication technologies are an integral part of the modern project of the sublime and its triumph over the pre-modern (Lyotard, 1989, p. 199).

The sublime is a persistent theme in the study of American culture and history, including such aspects as: the American sublime in nineteenth-century landscape writing and painting (Cole, 1835/1965; Furtwangler, 1993; Wilton and Barringer, 2002); the city sublime (Brown, 1959); sublime science (Miller, 1966); the rhetoric of the technological sublime (Leo Marx, 1964); the technological sublime (Kasson, 1976/1999); the rhetoric of the electrical sublime. (Carey and Quirk, 1989); the American technological sublime (Nye, 1994); and, most recently, the digital sublime (Mosco, 2004). Although the railway, and to some extent the telegraph, figure prominently in some of these studies, other communication technologies are generally neglected, as are their military origins and uses. Mosco is the exception as he devotes a chapter to the telegraph, electrification, the telephone, radio and television, though he does not discuss their military aspects.

The republic of the United States of America was founded in war, the War of Independence from its imperial master. It is a military state from its inception. Kasson touches on the military origins and applications of much early American technology, applicable also to much subsequent technology and communication technologies. Not only, as Kasson (1976/1999, p. 12) puts it in inverted biblical terms, did 'the nation beat its plough-shares into swords,' it also beat them into words, into code and communication technologies, such as aerial photography, two-way radios, computers, and satellites, into command-communication-control systems. The military, as Cowan (1997, p. 249) argues in her social history of American technology, 'has always been an important sponsor [and beneficiary I would add] of technological change,' including the communications technology revolution of the nineteenth and twentieth centuries (p. 273).

Recently Rosalind Williams (2000) took up Marx and Engels' phrase in the title of an essay reflecting on the impact of the information revolution on the contemporary university and on the historian of technology. Although she does not discuss the phrase in any detail nor relate it to the sublime, she suggests that 'all that is solid melt into air' is quite an apt description, not only for technology itself for the historian of technology in the information revolution, but also for describing the historian him or herself, especially of that revolution as it is occurring. He/she seems to be transformed out of existence as he/she encounters an unstable entity impacting on perceptions of time and space even as s/he researches it. In response Williams calls for passionate histories of technologies grounded in political critique and for relational histories linked to other disciplines. Grounding the history of communication technologies in political critique and linking it to other disciplines are the central purposes of this book.

Railways, telegraphy, radio, photography, cinema, cars, television, satellites, and computers are all sublime technologies. They are also products of the process of sublimation (and reproduce it themselves). Communication technologies transform the solid, living matter of bodies into the airy, dead matter of data. Communication technologies, such as radio and telecommunication satellites, sublimate (or 'sublime') the solidities of the earth and terrestrial space into the gaseous heights of the electromagnetosphere and orbital extraterrestrial space (see Giblett, 1996, chapter 2).

Moreover, communication technologies transcend space and time by enabling messages to travel faster than transportation. They dislocate local place and tempo into global, placeless space and almost instant-aneous time. Bodies performing actions are transformed into cyborgs engaging in events. This process, and the drive from which it arises, is masculine. It involves mastery over nature inside and outside the body rather than mutuality with it (see Giblett, 2004).

The account of communications technologies that I give is genea-logical in that a particular communication technology arises out of a previous one, or ones: television arose out of radio; radio out of tele-graphy; telegraphy out of letter writing; the Internet and computers converging out of all of them. This genealogy of communication tech-nologies is part of what Foucault (1977, p. 31) calls 'the history of the present.' Spectrum, computers, telecommunications and the Internet are key aspects of the present that have long histories in their precursors. The history of the present also points forward to 'the history of the future' (for a critical survey see Carey with Quirk, 1989b). Outer space and its control, space power and the 'weaponisation' of space are the scenarios envisaged by the new futurists rather than a global, spatial commons owned by none and shared by all. This history of the future ends up being a repetition of the history of the past: exploration, conquest, colonization and exploitation. What happened on earth is repeated in the heavens.

The genealogy is presented chronologically as the characteristics and structure of each succeeding generation of communication technologies were established very early on and bequeathed to each subsequent communication technology. Rather than an information revolution during the 1990s, we had an information generation, an information explosion not only in terms of content and quantity but also in the forms to express it and technologies to deliver it. Rather than convergence being an innovation of the 1990s between information technology and telecommunications, convergence has been the norm since railways and telegraphy when lines for the latter ran alongside lines for the former, and the latter was used to control the former. Divergence has also been the norm since then too as telegraphy communicated faster than railway, the then fastest means of transportation.

The first nine chapters of the book are concerned with the history of the present arising from the past; the final two are devoted to the history of the present going into the future. The first chapter begins by considering communication as a technology. Communication is basically the transportation of messages. Communication technologies arose out of transportation: telegraphy out of railways. Railways were also a communication technology in that not only did they carry messengers and messages, but also their windows and their curving tracks produced the landscape as a panoramic tableau for a seated passenger. The car as a communication technology was to inherit and develop these features of the railway with the possibility of stopping to take in the view of pleasing prospects from designated vantage points.

With the development of the telegraph communication was separated from transportation. Communication could go faster than transportation. They were divergent. Communication became analogously the transportation of messages. It became possible to talk about communication *as* transportation. Communication was no longer just transportation. It became separated from a particular time and place, and could transcend time and space. But there was still a physical connection between sender and receiver, the telegraph wire.

With the development of radio, communication was no longer reliant on a visible and vulnerable physical connection between sender and receiver. Communication became invisible. It now became reliant on the transmission and reception of signals through the electromagnetosphere. It became *metaphorically* the transportation of messages. It was possible to talk about communication *as* transportation. Communication was no longer transportation. But communication was still limited in range. With the development of communication satellites in the

1960s communication became almost instantaneous between widely separated places. It transcended earthly time and place and became located in geostationary, extraterrestrial orbital space.

After considering railways and telegraphy in the second and third chapters, succeeding chapters go on to consider modern communication and transportation technologies such as photography, cinema, car, radio, television, satellites, and computers. Industrial capitalism, and its forces of urbanization and globalisation, produced these technologies. They shaped the modern world, including human perceptions of time and space, and human bodily and mental capacities, and reshaped the earth.

Communication technologies and their users (and uses) are in communication with (and are a part of) the ecosphere. This book is both a cultural and ecological history of communications technologies. Communication technologies colonised and nationalised, enclosed and privatised the global commons of the ecosphere, including the electromagnetosphere, semiosphere and orbital extraterrestrial space. This book concludes with a call for rezoning and republicising the global commons of the communication sphere so that public interests are protected and promoted in the era of dominant private global communication conglomerates.

Acknowledgements

The dedication of this book to Zoë Sofoulis is a token expression of an embarrassingly long-standing (over a decade in fact) debt of gratitude to her for the generosity with which she shares her ideas and work with others. I am heavily indebted to Zoë for the theoretical underpinning for this book and previous ones stemming entirely from her work. Of course, she bears no responsibility for what I have done with it and for what follows. Zoë was also the first person to alert me to Paul Virilio's work, which is highly pertinent to, and figures prominently in, the present book.

Major intellectual debts are owed to two other critical commentators on communication technologies: Walter Benjamin and Raymond Williams. Benjamin, Virilio and Williams – or BVW, as I call the trio for short – are a sort of high-powered 'demobile' (demotic mobile, demobilised marque). All three have written extensively and critically on communication technologies, or on a particular communication technology, but none of them wrote, or in Virilio's case have written, an overall account of communication technologies. The current book assembles, elaborates and develops their scattered comments and insights with cognate work.

This book was researched, written and taught entirely in the supportive environment of the School of Communications and Contemporary Arts at Edith Cowan University.

I am grateful to Professor Robyn Quin for employing me at Edith Cowan University when no one else would and to David McKie for his support in initiating the process. I am also grateful to Robyn and my senior colleagues Associate Professor Brian Shoesmith and Associate Professor Arshad Omari for their continued support. Other present and past colleagues and postgraduate students too numerous to list made suggestions, gave advice and referred me to useful sources for all of which I am grateful.

I am also grateful for the opportunity to have delivered earlier versions of four chapters as papers at the annual conference of the Australian and New Zealand Communications Association (ANZCA): Chapter 9 at the 1998 ANZCA Conference (first published in the *Australian Journal of Communication (AJC)*, 28 (1), 2001); Chapter 6 at the 1999 ANZCA Conference (first published in (*AJC*), 27 (2), 2000); Chapter 4 at the 2000

ANZCA Conference (first published in the proceedings of the ANZCA 2000 Conference); and Chapter 11 at the 2001 ANZCA Conference (first published in *Media International Australia (MIA)*, *102*, 2002). I am also grateful to Ros Petelin, editor of *AJC*, for permission to reprint these portions of Chapters 6 and 9 first published in *AJC*, and to Helen Wilson, editor of *MIA*, for permission to reprint the section of Chapter 11 first published in *MIA*.

1
Flow Along a Channel: Communication and Its Technologies

In 1945 the inventor of the concept of the biosphere outlined some recent developments:

> In the twentieth century, man [*sic*], for the first time in the history of the earth, knew and embraced the whole biosphere, completed the geographic map of the planet earth, and colonised its whole surface.
>
> (Vernadsksy, 1945, p. 8)

Yet Vernadsksy overlooked some crucial additions made earlier in the twentieth century to the spheres of 'man's' influence, such as when 'he' used the electromagnetosphere to communicate over vast distances by radio. These developments followed, and were made possible by, some crucial additions made in the nineteenth century. A century before, in 1847 Daniel Webster proclaimed to his contemporaries that 'we see the ocean navigated and the solid land traversed by steam power, and intelligence communicated by electricity' (cited in Marx, L., 1964, p. 214; 1988b, p. 191). Yet the railway and the telegraph not only reshaped the earth but also enhanced human bodily and mental capacities as radio and television and information technology later were to do too. Intelligence was not simply communicated by electricity and left untouched as if it were just so much goods and chattels to be transported; it was also affected in and by the process. Intelligence also affected the process as these inventions and technologies were the products of the human brain.

Communication and transportation

Communication and transportation technologies not only shaped and extended human physical and mental aptitudes and abilities. They were

1

also shaped and created by human mental and physical aptitudes and abilities. In the nineteenth century, Friedberg (1993, pp. 3–4) argues, 'machines that changed the measure of space and time (machines of mobility, including trains...and, later, automobiles...) changed the relation between sight and bodily movement'. These changed relations between space and time, sight and bodily movement were also contemporaneous with, and produced by, what Friedberg calls 'machines of visibility': 'coincident with the new mobilities produced by changes in transportation, photography brought with it a virtual gaze, one that brought the past to the present, the distant to the near, the miniscule to its enlargement'. The scale, extent and duration of things were altered and perceptions of time and space changed forever.

The machines of mobility combined with the machines of visibility to produce what Friedberg calls 'machines of virtual transport (the panorama, the diorama, and later, the cinema [and still later radio, television and telecommunications which]) extended the virtual gaze of photography to provide virtual mobility'. The panorama was developed in 1804 by Robert Barker and was made up of a huge, circular painting (often 20 feet high by several hundred feet long) mounted on rollers inside a rotunda. Spectators were seated in the centre of the rotunda and separated from the painting by what Kattelle (2000, p. 15) calls 'a sort of dry moat' that removed and protected the audience from the scene, and mediated it to them. The panoramic painting 'scrolled past the audience', as Kattelle (2000, p. 15) puts it, in contrast to the diorama developed in 1822 by Daguerre in which the huge painting, as Kattelle goes on to describe, was

> now stationary with the audience [seated] on a rotating platform which carried them slowly past the varied scenes spread out before them...The foreground of the scene might contain actual objects, skilfully placed against the backdrop to give a strong three-dimensional effect.

The panorama and the diorama set the scene and developed or reproduced the viewing dynamics of transport and communication technologies. The panorama is to the cinema and television as the diorama is to the railway and the car. In the former, the depicted scene moves and the spectator is stationary; in the latter, the depicted scene is stationary and the spectator moves. Either way, in both cases, the scene moves, or appears to move, and both are vehicles for seated passengers. Both are also prosthetic devices that increase the visual capabilities and virtual range of the viewer. Both vehicles produce a sense of virtual mobility, virtual transport and the virtual gaze, all of which constitute the virtual world of hypermodernity stripped of a definitive object, of local space–time coordinates and of a solid sense of reality. They produce the world of sublime communication (and transportation) technologies.

Communication and transportation are inextricably linked and mutually supportive, as Cubitt (1998, p. 31) concurs with Freidberg.

> The histories of the photomechanical and electronic arts could be read as the narrative of assimilation and internalization of a combined communications (transport and mediation) apparatus in an ever-large urban population. The move from the machine ensemble of rail [combining track, engine, carriages, telegraphy, timetable and panoramic views (as we shall in the next chapter)] through the cinematic apparatus to virtual reality would then be just a hop, skip and a jump. This smoothness belies the insecurity of the transition.

Yet the transition was never really in doubt, and although there were some hesitations from hop to step to jump, there was a certain inevitability about it, especially in the way in which each succeeding generation of transportation and communication technologies combined and superseded the features of all the preceding ones. Convergence between communication technologies is not a recent phenomenon but the norm.

Continuing this line of thinking, a more extensive historical account than Vernadsky's at the beginning of this chapter that included the nineteenth century and the second half of the twentieth century could thus be devised along the following lines: in the mid-nineteenth century modernity colonised and enclosed terrestrial space and time by using the railway and telegraph to communicate via a physical link between vastly separated places. In the mid- and late nineteenth century it constructed vehicles of visualisation and virtual mobility for seated passengers that represented and enclosed visually landscapes in the panorama, the diorama and the cinema. In the late nineteenth century it colonised and enclosed the entire surface, and heights, of the earth by using radio to communicate across vast terrestrial distances via the electromagnetosphere.

Furthermore, in the early twentieth century modernity was hyper-masculinised by being militarised (Spretnak 1997, pp. 221 and 222). With the First World War, war was industrialised and mediated by rapid-firing guns, telegraphy and telephony. Furthermore, it was totalised and represented by aerial photography. At about the same time modernity was also mobilised in the private sphere of the car that provided actual mobility for seated passengers along highways and through landscapes constructed along panoramic lines. Following the First World War rocketry was developed and then used in the Second World War with deadly effect (Michaelis, 1978, pp. 859–865). Rockets were then used later in the 1950s and 1960s to launch satellites in the Cold War.

Since the Second World War hypermasculine modernity has further extended its reach beyond the biosphere and the electromagnetosphere into extraterrestrial space undertaking what Virilio (1994a, p. 201) describes as 'the conquest of sidereal space throughout the sixties'. In this decade it took what Smith calls 'the simulation of territorial and scientific conquest' (quoted by Wilson, 1992, p. 287) to new heights (literally) by exploring the space frontier, colonising orbital extraterrestrial space, enclosing the heavenly global commons and using it for terrestrial military purposes. In the late 1950s and early 1960s hypermasculine modernity placed satellites in orbit with the ability to 'penetrate' inhospitable terrain and to communicate between even more vastly separated terrestrial places than radio could alone.

In the 1970s orbital extraterrestrial space was militarised, with the majority of communication satellites being launched for military purposes. In the early 1990s the first war was fought using communication satellites stationed in orbital extraterrestrial space. The Gulf War of 1991 saw the coalition forces led by the United States rely not only on satellites for surveillance but also on computers for pinpoint targeting. War was taken to new heights with the development of extraterrestrial space as a new front, and frontier, and with computers as a new targeting technology. In the 1990s colonisation and exploitation of the electromagnetospheric spectrum took place as this new colony was opened up for 'blue sky mining' with the same boom or bust mentality that characterised gold rushes and railway speculation a hundred years before, and iron ore and nickel thirty years before. In the twenty-first century space is the new frontier for exploration and 'weaponisation'.

The boundaries of the ecosphere, the earth-household, the sphere in which we live and on which we depend, have thus been extended beyond those of the biosphere to take in the electromagnetosphere (including the ionosphere about 1000 kilometres above the earth used for long-distance radio transmissions) and orbital extraterrestrial space (with the geosynchronous orbit 35,900 kilometres (22,300 miles) above the equator). In the twentieth century hypermasculine modernity has embraced and known (in both the epistemological and the metaphorically sexual sense) the ecosphere (including the biosphere and the electromagnetosphere); penetrated inhospitable terrestrial space; mapped sidereal space; colonised and enclosed orbital extraterrestrial space; and continued the colonisation of the surfaces, depth and heights of the earth – all through the use of modern communication and transportation technologies.

This development can be traced further in terms of the spheres. The public sphere of the *polis* – the public sphere of the classical Greek city-state and later the bourgeois public (and private) spheres – colonised, enclosed, militarised and privatised the '-ospheres' (atmosphere, stratosphere, hydrosphere, lithosphere and biosphere (see Giblett, 2004, chapter 2)). Yet this militarisation of the -ospheres was part and parcel of the project of the *polis* in the first place. Indeed, the city, for Virilio (2002a, p. 5), 'the *polis*, is constitutive of the conflict called WAR, just as war is itself constitutive of the political form called the CITY'. The city and war are mutually constitutive, as Virilio and Lotringer (1983, p. 3) have argued extensively and as I have argued in relation to my own city, Perth, Western Australia (see Giblett, 1996, chapter 3).

Modern communication and transportation technologies both militarise and urbanise the surfaces, depth and heights of the earth. The urbanisation of the depths, heights and surfaces of the earth through what Serres (see Giblett, 2004, pp. 36–37) calls 'the plaque of suburbs and cityscapes' is just as much a process of militarisation as the militarisation of the depths in bunkers and shelters, the surfaces in planes, tanks and ships, and the heights in satellites and missile shields above the earth is a process of urbanisation (see Chapters 9 and 11). Mining the surfaces, depth and heights of the earth for materials, minerals and water is what the city and war depend on and what they consume greedily and parasitically, if not orally sadistically (see Giblett, 2004, chapter 9).

Communication technologies, too, still operate in the environment (in the broad sense) and are dependent on it however much they may wish to deny or repress it. Not only, as Jagtenberg and McKie (1997, p. 2) argue, is 'all communication ... biospheric in its action', but also all our actions are in communication with the -ospheres (as I argue elsewhere (see Giblett, 2004, chapter 2)). All communication takes place in the ecosphere and is dependent on its natural 'resources'; all communication technologies depend on, and exploit, natural resources. The communication frequencies that constitute the radio spectrum, for instance as Levin (1971, p. 1) puts it, are a natural resource essential to living in the modern world...Television and radio broadcasts [and even transmissions to point destinations] are as much a part of modern life as the air man [*sic*] breathes.

Television and radio, and computers and the Internet, are even more a part of modern life than the air we breathe, which we take for granted when it is good or ignore except when it is bad. If we no longer have roots but aerials, as McKenzie Wark (1994, p. 121) puts it, we still need air to breathe to live. During speaking engagements David Suzuki has asked members of his audience to see how long they can hold their breath in order to demonstrate that we are in symbiosis with the oxygen-producing plants of this planet. We are not only cyborgs, we are also symbionts, as Haraway (1985, 1995) puts it. We are interconnected not only with communication machines but also with the earth, its plants, animals, air and water. If we no longer have origins but terminals, as Wark (1994, p. 121) also puts it, we have a biological origin and term of, and termination to, our natural life.

Communication can be defined simply as the transportation of messages. Communication has not only been associated with transportation, but communication was synonymous with transportation until the development of telegraphy. Communication was transportation (full stop). Lines of communication between cities and countries, city and country were roads, rivers, canals, railway lines and shipping lanes (see McLuhan, 1964, p. 89; Williams, 1968, p. 17; 1976, p. 62). President Thomas Jefferson (cited in Dadley, 1996, p. 83) in his Sixth Annual Message of 1806 argued that 'the "channels of communication" created by a road and canal network would cement the union and make "lines of separation disappear"'. All that they did though was make lines of separation reappear between upper and lower classes, between North and South, industrial and feudal, east coast and west coast, city, suburb and country, frontier and wilderness.

Communication not only took place on these lines and in these channels but also on them communication increasingly sped up and increased in volume. Higher speed and increased volume are both features of the sublime. For Taylor (1973, p. 438), 'the rapid-speed up of communication and the great build-up of the volume of traffic brought the Sublime to the cities'. Speed is not only a quintessential feature of cities and modernity in general, as Virilio might argue, but also of the sublime in particular as the rush of sensory experience it gives rise to transcends a time, a place and the body. Smooth lines of communication enable sublime speed whether via road, rail or canal. All three fill up hollows and smooth over the surface of the earth to create gradients suitable for their means of transportation. Although canals seem passé (and the barge very slow) compared to steam railways and stage-coaches, they were, for Fishlock (2004, p. 14), 'far more efficient for bulk transport than roads'. Canals were what Taylor (1973, p. 438) calls 'the other weighty new arm of communication' of early modern industrialism to place alongside steam railways. Just as the steam railway was the crude conqueror of nature (as I will argue in the next chapter), the Panama Canal constructed in 1912 has been seen as 'the greatest liberty man has ever taken with nature' (Bryce cited in Fishlock, 2004, pp. 336 and 337). Liberties with nature were taken with this and other canals. Canals were not only the takers of liberties with nature but also the model for communication itself.

Sender→message→receiver

The transportation component or aspect of communication is implicit in the so-called 'process model of communication', in which a sender sends a message to a receiver. The model comes from transportation. Transportation and communication are now differentiated but it is useful to use transportation as a metaphor and as an explanatory, or heuristic, device for thinking about communication. Indeed, for Kress (1988, p. 4), 'the major model in communication theory could be taken to be a metaphor of nineteenth-century transport'. Or perhaps more precisely, nineteenth-century transportation could be taken to be a metaphor for nineteenth- and twentieth-century communication. Before transport became a metaphor for communication, communication *was* transport. There was what McLuhan (1964, p. 89) calls 'transportation as communication'.

Only when they were separated could transport become a metaphor for communication and could there be communication as transportation as well. Today, as Kress suggests, we can think of communication as transportation when, as Sofoulis (1990, p. 4) following McLuhan argues, 'every technology is a communication technology'. Elsewhere she argues, '*every technology is a reproductive technology*' (Sofia, 1984, p. 48, her emphasis) both in the sense that it reproduces and in the sense that it simulates reproduction. Every communication technology is a reproductive technology in both senses. These aspects of communication technologies will be developed in later chapters.

The product model of communication can be figured in terms of transportation: a merchant sends a bale of wool to a spinning factory.

The message is constituted as a thing or as thing-like in this model, as something that can be transported, as a product, as a package or packet. Communication is a process, 'the process of transmission and reception', as Raymond Williams (1968, p. 17) put it. Communication is a means of production (Williams, 1980b, pp. 50–63). Communication technologies, for Williams (1980b, p. 50), are 'always socially and materially produced, and ... reproduced'. But in the communication model the message is a product rather than a process and the process takes place outside the message. The message is simply something that the process does something to. The process is in the arrows in this symbolic formulation of the model: sender→message→receiver. The product model of communication leaves out the process of the production of the message itself and concentrates on the process of getting the message from sender to receiver, on getting the bale of wool from merchant to factory.

Yet a message is not like a bale of wool. It is not a product but a point in a process, a station in a progress. A message has meaning. It is not homogeneous but is comprised, like a bale of wool, of many strands and threads of meaning. A message is a visual image or a verbal text. Hence the power of the concept of message that it encompasses both visual image (or aural image in the case of music or speech) and verbal text. A text is made up of words, sentences and paragraphs, a visual image of pictures, bits, pixels and so on. A text is made up of signs, and a sign is a unit of meaning-making. A sign, in turn in Saussure's (1974) terms, is made up of a signifier (an aural image of a spoken word) and a signified (a concept). A visual signifier can signify a verbal signified. To make meaning, a visual signifier has to signify a verbal signified.

Here the analogy of communication with transportation ultimately breaks down as signs are not like strands or threads of wool. Signs can both refer to things outside themselves and produce meaning within themselves as it were. Words communicate ideas and 'stand also for the reality of things', as Locke (1947, p. 205) pointed out in the seventeenth century. Signs both signify and refer. For Locke 'men' communicate thoughts, or what he calls 'invisible ideas', and use what he calls 'external sensible signs' to do so to others (p. 204). We communicate signifieds (what Locke calls 'invisible ideas', or what Saussure calls 'concepts') and we use signifiers (what Locke calls 'external sensible signs' or what Saussure calls 'images') to do so.

For Locke there is 'no natural connection ... between particular articulate sounds and certain ideas' (p. 204). The relationship between sounds and ideas is arbitrary and conventional as Saussure was later to re-affirm (though Saussure excluded from structural linguistics the study of the relationship between signs and things). Between sounds and ideas is what Locke called a 'voluntary and perfect arbitrary imposition'. Words are 'voluntary signs'. The components of the message cease to be readily transportable. The components of the message are signifiable. They convey meaning. But the relationship between signifier and signified is arbitrary, contingent, conventional and even, in Voloshinov's terms, a site of struggle. The sign, for Voloshinov, is multiaccentual (1973, p. 23). A sign is not composed of just one signified; a signifier can signify many signifieds; for example, 'bicentennial' in Australia or France or the United States, or 'reconciliation' between indigenous and non-indigenous peoples as either a negotiated settlement between equals or one party reconciling itself to the facts of history imposed by another party.

This disjunction between signs and things is historically and culturally contingent. Signs until the seventeenth century, Foucault (1970, p. 129) argues, were 'part of things themselves [as they still are for traditional indigenous Australians], whereas in the seventeenth century they became modes of representation'. Locke is on the cusp of this shift as signs for him both refer to things and are modes of representation, whereas for Saussure in the twentieth century signs are only modes of representation. For traditional indigenous Australian cultures, rather than a sign referring to an object, an object is its sign; instead of living in capitalist modernity where signs are disconnected from their object, signs are their objects (see Giblett, 2004, chapter 11).

Signs and things became disconnected with the development of what Foucault (pp. 135, 157–162) calls the 'discourse of nature' (natural history) involving observation (seeing linked spontaneously with saying). Nature, for Foucault, is posited only through what he calls 'the grid of denominations' without which 'it would remain mute and invisible' (p. 160). Nature is made to speak, but made to speak in predetermined ways, in certain channels of discourse (especially taxonomy). This grid of denomination 'finds its locus in the gap that is now opened up between things and words' (pp. 129–130). The gap between things and words produces representation no longer tied to the thing itself.

Representation floats free from things, and for Foucault 'things touch against the banks of discourse because they appear in the hollow space of representation' (p. 130). This was 'a new way of connecting things both to the eye and to discourse' (p. 131). This was also a new way of *dis*connecting words and things. Things are disconnected from words and allowed to float free. Discourse only touches things. It only engages with them via the sense of touch and sight. It is never the things themselves. Things are the flotsam and jetsam in the channel of discourse. They bob along in the liquid flow of communication that has ceased to connect words and things, signs and referents, and become mere representation, a hollow space, the canal drained of significance, and reference.

Out of the channel of discourse things are plucked to become commodities. The signified and the referent, Baudrillard (1975, pp. 128–129) argues,

> are abolished to the sole profit of the play of signifiers, of a gener-
> alized formalization in which the code no longer refers back to any
> subjective or objective 'reality,' but to its own logic. The signifier
> becomes its own referent and the use value of sign disappears to the
> benefit of its ... exchange value alone. The sign no longer designates
> anything at all ... it refers back only to other signs. All reality then
> becomes the place of a ... structural simulation.

The sign becomes a commodity, and the commodity becomes a sign to the point that is impossible to separate them. Baudrillard (1981, pp. 147–148) argues later that 'nothing produced or exchanged today (objects, services, bodies, sex, culture, knowledge, etc.) can be decoded exclusively as a sign, nor solely measured as a commodity ... but [is] indissolubly both'. This disconnection between sign and things gave rise to, and made possible, the new technologies of representation in photography, cinematography, television and video in which signs are disconnected from things. These technologies became a hollow space of representation, a channel or canal for the transmission, or transportation, of messages.

The 'consumer' is the site of the intersection of both sign and thing. He/she consumes commodities as objects and as signs. The metaphor of the consumer is drawn 'from the stomach or the furnace', as Raymond Williams (1980a, p. 187) points out. The consumer is metaphorically a biological digestive organ and a thermodynamic machine for breaking down, melting, processing and reconstituting matter and signs, matter as signs. In Williams' (1980a, p. 187) terms, 'we are the channels along which the product flows and disappears'. We consumers, we 'ordinary members of modern capitalist society', as Williams (1980a, p. 187) calls us, do not sit on the banks of the channels of communication and watch things flow by. We are not outside the communication process, mere senders and receivers at one end of the process or the other. We *are* the channels. Signs and things flow through us.

Noise

Yet the message and the bale of wool may not reach its destination – it might get stolen, or fall off the barge. It will certainly not reach its destination in exactly the same condition as it left – it and its contents will have aged, the bale may get ripped and some of the wool may be lost or stolen. To explain the fact that the message received is never exactly the same as the message sent, that there is interference along the way, Shannon and Weaver, the inventors of the sender→message→receiver model, developed the concept-metaphor of noise (see Weaver, 1949, pp. 7–8, 18–22 and 26; see also Lax, 1997, p. 54). Noise in engineering terms is interference in the communication channel. Noise, for Weaver (1949, p. 8), 'may be distortions of sound (in telephony, for example) or static (in radio), or distortions in shape of shading of picture (television), or errors in transmission (telegraphy or facsimile), etc.'

Yet noise is not only interference, distortions or errors in the hardware but also anything that distracts or disables the receiver from receiving the message that is sent by the sender. Weaver (1949, p. 26) distinguishes between 'engineering noise', which impinges on the channel and which I associate with white noise (sterile, empty), and 'semantic noise', which impinges on the sender and receiver and which I associate with black noise (fecund, full). The latter for him is inserted between the information source and the transmitter, or between the pretext and the sender. He defines it as 'the perturbations or distortions which are not intended by the source but which inescapably affect the destination' (p. 26). The merchant makes a mistake in calculating the quantity of wool or number of bales he sends. At the receiving end of the process too semantic noise interferes in decoding the message and so affects the outcome. The factory owner makes a mistake in calculating the quantity of wool or number of bales he requires.

Yet without the possibility of making a mistake, the transaction would not be possible. Black noise makes communication possible. No noise, no communication. Even Weaver (1949, p. 19; see also p. 20) concedes that

> If noise is introduced, then the received message contains certain distortion, certain errors, certain extraneous material, that would certainly lead one to say that the received message exhibits, because of the effect of the noise, an increased uncertainty. But if the uncertainty is increased, the information is increased, and this sounds as though the noise were beneficial!

And indeed it is, but it is also detrimental. It is creative and destructive. Noise, for Serres (1982b, p. 12) too, is 'part of communication'. There is no communication without noise. The channel for him 'carries the flow' and there is 'no canal without noise' (Serres, 1982b, p. 79). There is no space in which a canal could be constructed without noise. For him 'there is no space without noise nor any noise without space' (Serres, 1995a, p. 79).

Yet noise is split between engineering (or white noise) and semantic (or black noise) (which Serres does not distinguish, or pick up from Weaver's work). The latter is an open space, or simply space, whereas the former is a line or vector through that (open) space. No communic-ation without a direction for communication to travel in (vector), and no communication without resistance to it (black noise) for it to travel through (white noise). White noise is the direction which communica-tion takes, or the line communication follows. For Serres (1995a, p. 58), 'noise is the vector'. Or more precisely, engineering noise is the vector. Without the vector of noise the communication channel can never be constructed; without the vector of noise that connects points A and B the communication canal can never connect them and communication can never occur between them. The vector is one-way as in the communic-ation model. It is monologic, not dialogic. It is the arrows in the model.

Noise is not only the vector for communication but also its milieu. Background noise, Serres (1995a, p. 7) argues, is 'the ground of our perception...the element of the software of our logic. It is the residue and the cesspool of our messages.' Black noise is the residual cesspool off the main channel of communication. It is the place where used and discarded messages rot. It is dirt in Mary Douglas's sense of matter out of place (see Douglas, 1966). Yet without black noise and dirt there would be no matter (and messages) in place, no clean and proper place for matter and messages, and no communication. Black noise and dirt are liminal elements that cross the boundaries between liquid and solid, matter and waste, message and static, meaning and nonsense – and make both possible.

Black noise occurs in both speech and writing. Cacophony is the black noise of the spoken form, whereas what Serres (1982a, p. 66) calls 'cacography' is 'the noise of the graphic form'. Cacography is bad spelling, and without bad spelling there is no good spelling. For Serres (1982a, p. 68), 'the act of eliminating cacography, the attempt to elim-inate noise, is at the same time the condition of the apprehension of the abstract form and the condition of the success of communication'. Yet there is no main channel of communication without this pool of rotting messages which not only interfere with all new messages but also provide the manure (and milieu) for them to grow in. No channels of communication without cesspools of noise dotted along the way.

Noise is an ambiguous substance, a mediator, the medium. The medium is not the message, the medium is noise. Black noise is the medium through which communication passes or travels and white noise is the vector passing through it. Black noise is neither solid nor liquid nor gaseous. Noise, for Serres (1995a, p. 13), 'is part of the in-itself, part of the for-itself'. Black noise is part solid and part liquid. Black noise is slime, neither solid nor liquid, both solid and liquid (Giblett, 1996). Black noise is not sublime, not gaseous. The sublime is solid becoming gaseous; black noise is solid becoming liquid. Black noise is the unspoken; the spoken sign is sublime. Speech is sublime, speech is breath, gaseous. Black noise is the trace, the ur-writing, that makes speech and inscriptive writing possible. No slimy black noise, no sublime communication and no sublime communication technologies either.

If, as Serres (1995a, p. 13) argues, 'background noise may well be the ground of our being', then the ground of our being is shaky, the place where the earth trembles as Okefenokee Swamp is called, what I have called elsewhere 'the quaking zone' (see Giblett, 1996). This is the place where nature begins. For Serres (1995a, p. 34), too, 'nascent nature begins in noise'. The birth of life begins in what Serres (following Michelet 1995a, p. 72; 1982a, p. 30), calls 'the prebiotic soup', or 'the primal liquid state', what Sartre calls 'slime' (see Giblett, 1996, chapter 2). Primeval slime is black noise. Space is born from black noise, from infinite temporality, eternity. Black noise is infinite temporality, whereas the sublime is infinite spatiality (see Giblett, 1996).

No communication without noise, no meaning without noise, no words without noise. Background noise, for Serres (1982a, p. 66), is '*essential* to communication' and for him there is 'no logos without noise...Noise is the background of information, the material of that form' (Serres, 1995a, p. 7). Information is the form, noise is the content. Both are produced out of the heat of noise and the making of meaning. For Serres (1982a, p. xxiv), 'information theory follows directly from thermodynamics', which Weaver (1949, p. 3, n.1 and p. 12) acknowledged happily anyway. Why? Because, for Serres (1982b, p. 169), 'there is only human production by fire and sign. By energy and information. Matter is energy; its form is information.' 'Pure information', for Zuboff (1988, p. 349), is 'light without heat'. It is the sunlight of reason without the heat. Information is cool in a number of senses including McLuhan's and more generally (see Barthes, 1979, pp. 43–45).

Matter is black noise, the material on which heat and energy are exercised to transform black noise into information, to make meaning out of meaninglessness, to give form, not to the formless or informal, but to the aformal, that which is outside and before form, primeval slime. The cesspool generates heat through decomposition of matter. Out of the application of heat to dead matter new living matter is generated. All cultures and societies have done this but modern societies have then gone on to transform solid living beings into gaseous dead matter through the hot, thermodynamic technologies of industrial capitalism and the cool communication technologies of informational capitalism. Solid matter is transformed into gas in the process of the sublime.

The sublime is one of the most prominent and striking aspects of modernity, not only of modern aesthetics but also of modern culture in general, including aesthetics and technology (see Giblett, 1996, chapter 2). Sublimating communication takes it out of transportation into telecommunication. It takes it out of the sphere of face-to-face interaction into the electromagnetosphere – into gas. It also extracts information out of what Serres (1995a, pp. 32 and 44) describes as 'blank meaning', or 'what Plato named the *chôra*, a smooth and blank space prior to the sign'.

Onto this screen of blank meaning the figure of information is projected. Benjamin (1999b, p. 72) describes how 'the youthful goddess of information performs her provocative belly dance'. Information theory then comes along and cleans up her act. For Weaver (1949, p. 27), 'an engineering communication theory is just like a very proper and discreet girl accepting your telegram'. The discreet office girl of engineering theory makes respectable the belly dance of the provocative youthful goddess of information.

Information is a gendered category; information and communication theories are aspects of the gendered construction of reality. The youthful, belly-dancing goddess of commodified signification is later photographed and cinematographed to become a star fixed in the sublimated realm of heavenly bodies in pictorial advertising and cinematic pictures. She is the cool, smooth surface or screen of the body of the beautiful young woman onto which masculine phantasies are projected. In the process she becomes a dead living star who continues to shine brightly in our secular heaven however long she may be dead or however long she may continue to live.

Information theory, Serres (1982a, p. 73) argues, 'was considered the daughter of thermodynamics'. Information is the ice-maiden daughter of the thermodynamic technologies of communication that used heat and light. Black noise, on the other hand, is the Great Goddess, the black and slimy depths of the body of the mother. 'The noise–information couple', as Serres (1982a, p. 80) calls it, the mother–daughter dyad, the Goddess–Star relationship, is the feminised figure of pre-modernity and modern communication technologies combined.

Black noise is primal without being primary. Noise, for Serres (1995a, p. 54), is 'before language, before even the word'. In the beginning was *not* the word (contrary to the biblical gospel of St John, chapter 1, verse 1), nor in the beginning was black noise (and contrary to what Serres (1982b, p. 13) thought previously when he proclaimed that 'in the beginning was the noise'). Nor in the beginning was the Big Bang as physics thinks (Serres, 1995a, p. 61). Black noise is primeval slime, but it was not there in the beginning; it is not primary. In the beginning, for Buber (1970, pp. 69 and 78), is 'the relation'. In the beginning, specifically for Serres (1995, p. 70), 'is the echo: murmur' and 'in the beginning is the song' (Serres, 1995, p. 138). The song is a performance of the body that leaves traces; it is not an inscription that makes marks on a surface. First song: mother to child. First echo, first murmur: child to mother.

Writing

Communication as performance or process and communication as inscription or production are two extremes or poles between which communication takes place. Both depend on black noise. Whereas one acknowledges and thrives on black noise, the other denies and represses it. Both use or are constituted around different concepts and practices of writing. All people write in the sense of making meaningful marks on surfaces such as on the body (tattooing, painting) or on the earth (songlines, rock painting, smoke signals, cities) or on (trans)portable objects (message sticks) or with the body (dancing, Tai Chi, yoga).

The crucial distinction for understanding cultural difference is not between orality and (alphabetic) literacy as all cultures have writing and all cultures are literate in reading signs and making meaning from the signs they produce. Rather the crucial distinction is between the trace and the inscription, between traditional, matrifocal cultures and patriarchal (including modern) ones; between cultures of first, or worked, nature and cultures of second, or worked-over (even over-worked), nature – and not between nature and culture (see Giblett, 2004, chapter 1).

For a new discipline or methodology that studies writing as trace, communication as process, black noise as necessary, Serres (1995a, p. 21) has invented the term 'ichnography' from the Greek word *'ichnos'* meaning 'the imprint of a foot, the trace of a step'. Perhaps Serres is thinking of Robinson Crusoe coming across a footprint of Friday in the sand, which not only indicates another human being, but also interferes with his settled, self-enclosed castaway existence. It constitutes both threat and opportunity. It is highly ambiguous; it is black noise. Ichnography, for Serres (1995a, p. 21), is 'noise itself'. On the seashore or in the slimy swamp of black noise the footprint is quickly filled up with water and disappears.

Writing as inscription is distinct from writing as trace (see Giblett, 1996, chapter 3). There are not two sorts of writing, but writing is double, or split between the desire to perform and to leave only traces, and the drive to inscribe and to leave durable, if not permanent, marks, to erect monuments. The body is a site of performance of traces and a surface of inscription where the body is not only the human body but also the body of the earth.

Technology of communications

Communication technologies do not simply transport messages but also transform messages, messengers and receivers. The communication model shows that communication is a process and that communication can be considered as transportation. Communication is also a technology just as much as any of the machines of communication. It involves aptitudes, skills and training. Any consideration of communication technologies should begin with what Serres (1982a, p. 125) calls 'the technology of communications', not only the myriad acts of communication but also communication *as* a technology. Considering communication as transportation also means that communication itself can be considered as technology.

Technology is not just, or even, the hardware but also the interface and interaction between software, hardware and wetware (humans and other living beings). In 1829 Bigelow coined the word 'technology' in *Elements of technology*. This book, for Perry Miller (1966, p. 289), 'should be honored as a major document in American intellectual development'. Bigelow defined technology as 'the principles, processes and nomenclatures of the more conspicuous arts, particularly those which may be considered useful' (cited in Dadley, 1996, p. 82 and Uglow, 1996, p. 3; see also Cowan, 1997, p. 204 and Noble, 1999, p. 83). One of the uses of the term 'technology', for Bigelow (cited in Miller, 1966, p. 289 and Nye, 1994, p. 45), was couched in the unashamedly sublime terms that 'we ascend above the clouds' and 'extend the dominion of man over nature'. Besides being recruited for the trajectory of the sublime and subsumed to the utilitarian principle, technology is what Bigelow (cited in Noble, 1999, p. 93) called the 'arts of science'.

In more recent terms, technology is the conjunction of theory, knowledge, discourse, power and practice. Technology is an institutionalised practice of meaning-making before it is the mechanical production of machines and the use of those machines. A technology, for Raymond Williams (1981, p. 226), is 'first, the body of knowledge appropriate to the development of technical skills [or techniques] and applications [in a technical invention] and second, a body of knowledge and conditions for the practical use and application for a range of devices'. Technology is the coordination of brain, hand and eye. In displacing the other senses and organs, and in moving the emphasis upwards from the lower to the upper senses and organs, technology is a sublimation of the body. In turn, and in Brown's (1959, p. 297) terms, 'technological progress makes increased sublimation possible'.

By transcending local time and place and creating a virtual world, modern transportation and communication technologies were at the forefront of the sublime project of modernity. Modern transportation and communication technologies were not merely couched, nor were they simply bruited, in (nor did Daniel Webster in 1847 merely employ as a stylistic flourish) what Leo Marx (1964) calls 'the rhetoric of the technological sublime'. Rather, these technologies *were* sublime – they enacted the sublime. They transformed the solidities of traditional societies with their stable and settled sense of local place and seasonal rhythms of time into the gaseous nothingness of modernity with its sense of uprooted and restless placelessness and near instantaneous timelessness.

Yet the sublime is always haunted by its 'other', which is, as I have argued elsewhere, following Sofoulis, slime (see Giblett, 1996, chapter 2). Slime, Sofoulis argues, is the secret of the sublime which she encapsulates in the parenthetical portmanteau s(ub)lime (see Giblett, 1996, chapter 2). Sublime communication technologies not only dislocate and distemper a grasp of local time and place but also produce an uncanny sense of what Jeremy Sconce (2000) calls 'electronic presence'. Communication technologies, especially as used in the mass media, not only 'kill' people and places by representing them in the dead matter of the visual and aural image, but also bring them back to life, or at least give them a strange kind of half-spectral afterlife in the flickering shadows, and shades, of the same medium. What Sconce (2000, pp. 3 and 12) calls 'the media's awesome power of animated "living" presence' produces 'either the beatitude of an electronically liberated subject or the incarcerating mirages of an encroaching electronic subjectivity'. Or both.

Rather than a sterile antinomy that situates proponents of the media and opponents of them on either side of a divide of a mutually exclusive antagonism of either liberation or incarceration, this book argues that the media do both. It acknowledges, for instance, not only, as Sconce (2000, p. 81) puts it, 'radio's uncanny liberation of the body in time and space', but also radio's unfortunate incarceration of the body in home and car. Or perhaps more precisely, it recognises radio's simultaneous and concomitant uncanny and sublime liberation and incarceration of the body. Sublime communication technologies are also uncanny communication technologies. The uncanny, for Freud, was a return *to* the repressed brought about through ghostly, virtual presence evoked by crafted artefacts coming alive as it were, and through the disturbing simulation of human movement and speech by automatons (see Giblett, 1996, chapter 2). A more apt description of television has probably never been written!

Uncanny communication technologies combine both aspects as they are both manufactured commodities and prosthetic devices. Communication technologies promise sublimation of the body into the mind, but technologies never keep their promise, never deliver on their claims, or more precisely on the promise and claims made for them by their proponents. Instead of transcending the body, time and space, the uncanny electronic presence of the media returns the listener and the viewer to their own body, time and place eventually. The upward displacement of the sublime is countered by the downward emplacement of the uncanny (see Giblett, 1996, chapter 2, especially Figure 1). Radio and television, the cinema, the car and the Internet liberate the mind from local time and place and sublimate the body into timelessness and space at the same time as they incarcerate the body in an interior space in front of a radio or television set, or the cinema screen or computer screen, or behind a car windscreen or train window.

2
Crude Conqueror of Nature: Steam Railways

In 1874 Jasper F. Cropsey, a member of the 'Hudson River School' of landscape painters in the United States, painted a picture that he called *Sidney Plains with the Union of the Susquehanna and Unadilla Rivers*. The title notwithstanding, the painting is just as much about the railway and telegraph as it is about the two rivers. The left foreground shows sheep safely grazing. The mid-foreground depicts a line of telegraph poles with a tenuous hold on the banks of one of the rivers. The rest of the foreground and the whole of the midground are dominated by the two rivers and the intervening land much of which is wetland. In the background a wisp of smoke rises from a steam train crossing the landscape. This rising smoke is 'echoed' or 'rhymed' in the smoke rising from the top, or over the other side, of the hills in the background. Whether this smoke is rising from a fire inside a friendly settler's cottage or from a threatening Amerindigenes' open fire is another question, though in this location at that time it is probably the former (for a reproduction, see Ellis 1998).

Either way, both columns of smoke are forms of telecommunication in the basic, etymological sense of 'communication over long distance' (Livingston, 1996, p. 5). Yet only the smoke rising from the hill would carry much of a message as it did in early colonial Australia, where 'columns of rising smoke, with their alerting message of change, were the earliest telecommunications to be used in Australia' (Moyal, 1984, p. 1). They were the first 'bush telegraph' that complemented the 'bush telephone' of the bullroarer. Aboriginal smoke signals, according to Carroll (1992, p. 3), gave 'simple messages' conveying location of people or 'game' and did not seem to have 'any special code'. The smoke rising from the train in Cropsey's painting is even simpler as it is merely an index of its presence. The train carried messages, and messengers. It was also a message itself. Industrial and urban communication and culture were superseding the form of communication and culture (indigenous or yeoman settler) indicated by the smoke rising from the hill. The rising smoke is nicely ambiguous as it refers to pre-industrial communication and/or culture in general – both indigenous and settler.

Moreover, the railway and the telegraph are shown in the painting in two ways: first, as emblems of modernity superseding pre-industrial forms of communication and culture; and secondly, as tenuous figures subsumed aesthetically within the immensity of the landscape when in fact they were successfully subduing the land economically. The limited visibility of the railway and telegraph in this painting belied their dominance over the land; the dominant visibility of the land in the painting belied its powerlessness before the railway and the telegraph. The dominance of the land in the painting compensated for the dominance of the railway and telegraph over the land. It also consoled the viewer with its pastoral vision of farming and technology working in harmony. Cropsey's painting for Amy Ellis (1998, p. 172) is 'a vast panoramic depiction of a settled valley, where the pastoral ideal meets the technological: sheep graze on the hillside in the left foreground, and the Erie Railroad makes its way across the middle background at the foot of the hills'. Other symbols of the pastoral, or more precisely, of the agricultural, such as sheaves of grain and a fence, are also placed in the left foreground.

As a panoramic landscape painting, the viewer is positioned to 'read' the painting laterally, and chrono-topographically (and chrono-tropically), from left to right, from the pastoral to the technological, from the primitive to the modern in the onward march of progress. The viewer is also invited to participate in the modern mastery of the earth symbolised by the railway and the telegraph. Their lines are ruled across the surface of the landscape and the two rivers, and the intervening and surrounding wetland and dryland, are scrawled, and contained, between and by them. Similar sentiments were voiced in a poem penned to celebrate the completion of the Trans-Australian Railway in 1917 when a poet described, 'splashes of smoke and shining steel/On the face of nature scrawled' (cited in Burke, 1991, p. xiii). The railway wrote lines on the face of the earth just as Cropsey drew lines on canvas in his painting.

Cropsey's painting is a perfect illustration of the fact that the railroad is, in Slotkin's (1994, p. 214; see also Smith, 1950) terms, 'an enterprise that represented 'the industrial revolution incarnate, but which at the same time appeared to offer a benign and productive association between the order of industrialism and the ambitions of the yeoman farmer'. Farming, telegraphy and railway coexist peacefully, if not live together harmoniously in the painting, and were supposed to do so in the wider culture as well. In another painting, *Starrucca Viaduct*, in which the railway is also one of the main focuses, Cropsey depicts what Wilton (in Wilton and Barringer, 2002, p. 140) calls 'technological innovation in the form [of the railway as] a welcome and harmonious addition to the American landscape' (for a reproduction see Wilton and Barringer, 2002, p. 141).

These two paintings, like so many other illustrations of the period (see Novak, 1980; Danly, 1988; Marx, L., 1988b), narrate what Slotkin (1994, p. 215) calls

> a fable of history in which the remnant of wilderness, the plowed fields, and the steaming locomotive represent progressive stages of development. Just as the farm is superior to and replaces the wilderness, so too the railroad represents a further development of civilized power, which always take the form of establishing human control over nature. The building of railroads allows men to evade the limitations of nature.

Furthermore, it enables 'men' to sublimate solid matter into steam, to calculate themselves as independent of nature, to attain the dynamically sublime and to escape the earthly constraints of space and time 'just as smoke and steam dispel into thin air', as Novak (1980, p. 170) puts it, and just as 'all that is solid melts into air', as Marx and Engels put it. The railway was not merely couched in the rhetoric of the technological sublime, but *was* the technological sublime (as we will see in greater detail below).

Railways are a 'way in the wilderness' leading the children of God out of captivity and slavery into the promised land – both literally and figuratively. When Sir John Forrest turned the first sod of the Trans-Australian railway in Port Augusta, South Australia, in 1912 he quoted a verse from the bible in which God says, 'See, I am doing a new deed / even now it comes to sight; / can you not see it? / Yes, I am making a road / through the wilderness, / paths in the wild' (Isaiah 43:21). Burke (1991, p. i) not only alludes to this verse in the title of his history of this railway, but also uses it as an epigraph to his book and notes that Forrest quoted it on the occasion mentioned. Yet Burke does not acknowledge that Isaiah ascribes these words to God and that by appropriating such sentiments Forrest was performing the ultimate hubris, and Satanic sin, of likening himself to, or even being, God who creates matter out of nothing and whose first words in the bible are 'let there be light'.

The railway is the enactment of belief in the secular theology of progress and the bringer of light, and enlightenment, to the darkness. Such sentiments were mobilised in the service of the formation of nations. The Trans-Australian Railway played an important part in the creation of Australia as a nation. Forrest was one of the 'founding fathers' of Australian Federation and the east–west Trans-Australian Railway is considered by Burke (1991, p. viii) in the subtitle to his book to be 'the first great work of Australia's Federation'. Wild in his foreword to Burke's book (1991, p. xvi) goes even further to regard it as 'the most important project in the history of the Australian nation'. Forrest was not merely making a way in the wilderness and turning darkness into light like (or as) God, but was giving birth to the nation just as God created heaven and earth, and light, by fiat.

The way in the wilderness of the railway leading to the promised land of nationhood progresses in the nineteenth century in settler societies such as Australia and the United States and, in Cropsey's painting, to the next step in the triumphal march of modernity: the wires in the wilderness of the telegraph wiring a continent and nation together. The railway and the telegraph were not only evading the limitations of nature and forming modern, industrial nations but were also represented at the same time as being in harmony with nature. The pastoral cloaked and ameliorated the industrial. Cropsey's painting enacts what I have called elsewhere 'the pastoro-technical idyll' in picturesque mode in which modern, industrial technology and nature (or at least, the second, or worked, nature of pastoralism and agriculture) are shown to be living together organically in harmony (see Giblett, 2004, pp. 66–67).

This pastoro-technical idyll of 'the machine in the garden', in Leo Marx's (1964) terms, depicted in nineteenth-century landscape painting and poetry was to give way to, and be superseded by, what Berman (1983) calls 'the techno-pastoral ideal' of what Kasson (1976/1999, p. 165) calls 'the garden in the machine'. This was enacted in the panorama and by the *flâneur* in the nineteenth century, and espoused by Futurist manifestoes and constructed in modernist architecture such as le Corbusier's in the twentieth century. In the techno-pastoral ideal modern, industrial technology is given pastoral qualities of organic harmony and humans graze in its manufactured streetscape paddocks and are put to bed in apartment pens like so many docile sheep, just like the sheep grazing safely in the left foreground of Cropsey's pastoro-technical painting.

Rocket railways

Railways have an important place in modernity in the modification of the land to accommodate railway lines and in their association with, and use in, industrial capitalism, nation-building and modern warfare. Railways for the geographers O'Dell and Richards (1971, p. 17) 'represent a new stage in the history of civilization, a stage characterized by an increase in speed and power of the means of transport [and thereby of communication]. They have revolutionized geographical conditions on our planet.' They also revolutionised cultural conditions, including the conduct of war, and work. For Novak (1980, p. 166) they are 'the nineteenth century's most ruthless emblem of power'. Martin (1992, p. 12) agrees that 'there has never been any sustained attack on the idea that the steam railroad was the most significant invention or innovation in the rise of an industrial society'. He concludes that the railroad, 'and the telegraph, likewise, were true breaks with the past' (Martin, 1992, p. 31).

To appreciate this break and the immediate, contemporary impact of the railway and the telegraph on the land and on people's lives we need to look back on the time when, as Marvin (1988) puts it, the old technologies were new. Yet Lévy (2001, p. 3) objects to this typical usage of the word 'impact' as it is, as he points out, 'a ballistic metaphor' in which 'technology is compared to a projectile...and culture or society to a living target'. 'Impact' has also become 'the normal description of the effect of successful communication' as Raymond Williams (1980a, p. 190) argued. Lévy goes on to critique this usage as for him 'the impact metaphor' is inadequate in discussing 'the 'impact' of new information technology' (and presumably of communication) as for him this technology is not a projectile nor is its cultural or social target living.

Yet rather than transportation and communication technologies being projectiles that have an impact on their users as if they were outside them and they were their target, the technologies *themselves* are projectiles in which users travel in reality or virtually (and which have an 'impact' or effect on them for that reason). They also have an impact on the earth (rather than just on culture or society) as a living target. It is precisely in terms of a projectile that communications technologies beginning with the railway were conceptualised. The train for Schivelbusch (1986, pp. 54 and 129) is

> experienced as a projectile, and travelling on it, as being shot through the landscape...Thus the rails, cuttings, and tunnels appeared as the

barrel through which the projectile of the train passes ... The railroad train as a projectile shot through space and time

is both constructive and destructive. All communication technologies have a trajectory and a vector along which they, or communication, travel and through which they project the user; all of them reduce the depths of the earth to the surfaces of landscape; and all of them target the living body and earth and have a destructive 'impact' on them. Information technology is not all that different from transportation technology; both are communication technologies.

Actress Fanny Kemble figured the steam railway as projectile in 1830 in her description of a journey 'Aboard Stephenson's "Rocket"' (a projectile by name if there ever were one) built the year before (see Martin, 1992, p. 13):

We were introduced to the little engine which was to drag us along the rails ... This snorting little animal, which I felt rather inclined to pat, was then harnassed to our carriage ... The steam-horse being ill-adapted for going up and down hill, the road was kept at a certain level, and appeared sometimes to sink below the surface of the earth, and sometimes to rise above it. Almost at starting it was cut through the solid rock, which formed a wall on either side of it, about sixty feet high. You can't imagine how strange it seemed to be journeying on thus, without any visible cause of progress other than the magical machine, with its flying white breath and rhythmical unvarying pace ... You cannot conceive what that sensation of cutting the air was: the motion is as smooth as possible, too. I could either have read or written ... I stood up with my bonnet off ... The sensation of flying was delightful and strange beyond description.

(Nicholl, 1997, pp. 23, 24; Fishlock, 2004, p. 185)

What Fishlock (2004, p. 185) calls 'the sensation of skimming the earth like birds' was written about, and revelled in, by many early railway travellers such as Kemble. It was also indicative of the sublimatory power of the steam railway to take travellers above the earth in the first form of mechanical flight, and of extraterrestrial travel. After all, Kemble was travelling in a rocket.

The construction of railway lines necessitated a massive reconstruction of the land. Railways, for geographers O'Dell and Richards (1971, p. 17), 'form an integral part of the cultural landscape' which they define simply as 'the modification of the natural landscape by man' [*sic*]. The railway line cuts through stone so that the train can cut through the air. The transformation of stone into air is the process of the sublime whereby solid matter is transformed into gas. The train is a magical machine (the sublime is modern magic) for transforming solid matter (including organic matter) into steam. Railways are sublime (see Nye, 1994, pp. 45, 56).

The industrial machine is a sublime (and sublimated) animal. The train is a domesticated mechanical horse, the 'iron horse' of nineteenth-century American culture and twentieth-century American 'westerns' beginning with John Ford's film of the same name (see Marx, L., 1964, pp. 195, 202, 207 and 252; Kirby, 1997, p. 1). The 'iron horse', for Leo Marx (1988b, p. 191),

> as depicted in the more nationalistic celebrations of the technological sublime, became a divinely ordained instrument for penetrating the wilderness, driving out the Native Americans, subduing the earth, and taking dominion over the vast trans-Mississippi West.

The iron horse worked with the 'iron cord' of the telegraph line, but the latter went ahead of the former in the United States to reach the Pacific eight years before it did (Thompson, 1947, p. 204).

In Australia the singing line of the telegraph went from south to north over one hundred years ago, whereas the north–south railway line was completed in 2003. The east–west Trans-Australian railway provided an interstate and federating transportation link 45 years after the north–south telegraph provided a communication link between the centre of imperial power in London and the far-flung Australian colonies between which telegraphic communication had already been established. The railway line followed the tele-graph line trans-continentally over very long distances, whereas the opposite applied over shorter distances within and between cities and states.

Yet the mechanical animal of the railway was not only seen as a docile, domesticated servant. The railroad invoked the stock devices not only of the technological sublime but also of what Leo Marx (1964, p. 211) calls 'the monstrous machine'. This combination of the sublime and the monstrous encapsulates the paradox of modern technology: liberating and enslaving; empowering and disabling; consuming and devouring; constructive and destructive at the same time – in peace and in war (see Schivelbusch, 1986, pp. 129 and 158). Even in the same piece of writing one writer had no qualms about employing both devices. Railroads, for Lanman, in 1840 were not only 'the triumphs of our own age, the laurels of mechanical philosophy, of untrammelled mind, and a liberal commerce', but also the 'iron monsters' 'feeding upon wood and water... leaping forward like some black monster, upon its iron path, by the light of the fire and smoke which it vomits forth' (cited in Marx, L., 1964, p. 207; 1988a, p. 134). The railway is an oral sadistic and bulimic monster that consumes and expels half-digested waste through the same hole, or more precisely mouth. Perhaps it is hardly surprising that Amerindigenes of the same period also saw the railway as monster. They did not like 'the Iron Horse that ran on iron tracks, whistling and snorting smoke, and frightening all the game' away (Brown, 1970, p. 139).

The iron horse not only consumed sadistically but also reshaped the earth as it runs on iron rails laid in parallel lines. In the words of Henry Colman in 1838 (cited in Miller, 1966, p. 309), 'mountains [are] levelled and valleys filled'. The railway line cuts through the humps and fills up the hollows to make a smooth, planar space free of obstacles to the trajectory of travel. The railways are just one of the mechanic arts that, for the Reverend Mr Williamson speaking in 1841, 'change the face of nature'. They are thereby the means whereby 'the mountains are leveled – the crooked is made strait [sic] and the rough places plain' (cited in Marx, L., 1988a, p. 133). The railway fulfilled an evangelistic function by virtue of its construction or form irrespective of what, or who, it carried, its contents. The railway smooths out or fills up the hollow, low-lying places of the earth by bridging or filling them, or both; the railway colonises the land, including wetlands (see Giblett, 1996), transforming it into landscape, into aesthetic surface (see Giblett, 2004).

The destructive impact of the railway on the land was a common complaint in the nineteenth century. In 1838 Victor Considérant (cited in Benjamin, 1999a, p. 636) described how

> The operation of railroads... forced humanity into the position of combating nature's works everywhere on earth, of filling up valleys and breaching mountains, ... of struggling finally, by means of a general system, against the natural conditions of the planet's terrain, ... and replacing them *universally* by the opposite sort of conditions.

Railways necessitated the greater-than Sisyphean task of moving mountains, or at least breaching them. Whereas Sisyphus merely rolled a rock up a mountain only to have it roll back down again and then have to start again, successful super-Sisyphean railway engineers rolled back mountains – and there they stayed.

Yet, like Sisyphus, once they had finished there was always another mountain to roll back. And like Sisyphus who performed his task in Hades, super-Sisyphean railway engineers performed their task in the hell of modernity. This is a place of both combustion and excommunication, especially for colonial exiles in the past (Moyal, 1984, p. 1) and for the developing world in the present and foreseeable future (Cubitt, 1998, p. 149). The new technologies of communication perform this task because for Virilio (1995, p. 20) they 'endlessly aggravate the casting aside that excommunication used to accomplish plunging the greatest number into a socially untenable *reality effect* with all the resultant geopolitical chaos'. Technological and endocolonial communication and excommunication replace theological and colonial excommunication. Instead of being excommunicated by the church from the community of saints and the saved for various sins, one is now excommunicated from the community of the digerati by the sin of not having access to, or not using, various communication technologies.

Invoking Sisyphus to figure the Herculean labours of the railway engineer was part of a consistent mythologisation of industrial technology that took place in the nineteenth century and still persists in the twenty-first. Yet Karl Marx argued that ancient mythology could have nothing to do with modern technology as industrial capitalism and modern machinery had superseded the world of the Greek gods by practising materially what they had only symbolised imaginatively. In his 1857 'Introduction' to the *Grundrisse* (the economic manuscripts of 1857–58) Marx (1986, p. 47) asked,

Is the conception of nature and social relations which underlies Greek imagination and therefore Greek art possible in the age of selfactors, railways, locomotives and electric telegraphs? What is Vulcan compared with Roberts and Co., Jupiter compared with the lightning conductor, and Hermes compared with the Crédit Mobilier? All mythology subdues, dominates and fashions the forces of nature in the imagination and through the imagination; it therefore disappears when real domination over these is established. What becomes of Fama beside Printing House Square? Greek art presupposes Greek mythology, in other words, nature and even the social forms have already been worked up in an unconsciously artistic manner by the popular imagination... is Archilles possible when powder and shot have been invented? And is the *Iliad* possible at all when the printing press and even printing machines exist? Does not the press bar inevitably spell the end of singing and reciting and the muses, that is, do not the conditions necessary for epic poetry disappear?

According to Marx, the conception of nature and social relations which underpinned Greek imagination, and therefore Greek art, is not possible in the age of selfactors, railways, locomotives and electric telegraphs. Presumably it would not be possible too in the age of computers, cars, satellites and the Internet, yet Hermes has been invoked recently as the god of e-commerce. Vulcan, the Roman god of blacksmiths and the forge, may not rate much against the designer of tools, machines and locomotives and the inventor of the selfactor. But modern heavy and dirty industry needs a mythology to explain itself and a god to call its own and Vulcan is an obvious choice.

Mythology does not disappear when real domination over the forces of nature in and through the imagination is established. Industrial technology that actually dominates nature replaces pre-industrial mythology that imaginatively dominates it. Domination of nature goes on – just the means and the ability to do so changes as does the status and function of mythology from imaginative domination of nature before the fact to industrial domination, mythological reflection, narrative explanation and symbolic legitimation after the fact. The former is prospective in temporal orientation; the latter retrospective. Despite the claim of the latter to be forward-thinking and progressive, it is constantly turning back to the past for legitimation for its actions in the present and future, but in the process mythology ceases to be divine theology and becomes secular theology.

God, and the gods, may be dead, but their mythology lives on. The forces they embody are divested of divine power. Instead of representing forces, the forces represent the gods. The sacred becomes secular. Mythology and its divine forces and heroes are secularised. Archilles may not be possible when powder and shot have been invented but modern warfare needed a hero and Archilles would do. If it needed a god of war too, Mars would continue to be it. And if the *Iliad* is not at all possible when the printing press and even printing machines exist in the sense that the printing press killed oral poetry, the *Iliad* lives on in printed texts, a pale spectre of its former self.

Time and the sublime rhyme

Some early commentators on railways wanted to have a bet both ways and enumerate their ills whilst protesting that their object of critique was not the railway *per se* but only the *uses* to which it was put. This became a standard cavilling response to industrial technology in which it was regarded as neutral and only the uses to which it was put as detrimental or destructive. Yet a railway that is not used is a train sitting idle on an otherwise empty line. The railway is the use. A technology and the use to which it is put cannot be separated out – the technology is the use. It should come as no surprise that a cavilling response came from a Romantic poet who espoused the pastoro-technical idyll (see Giblett, 2004). In 1844 Wordsworth (1906, pp. 162–163) described the railway aptly as 'Pests' with 'its scarifications, its intersections, its noisy machinery, its smoke and swarms of pleasure-hunters'. Railways scar the living surface of the earth in three dimensions with their lines and grids going across and cutting into the land.

Yet Wordsworth (1906, p. 164) also wanted to have a bet the other way:

Once for all let me declare that it is not against Railways but against the abuse of them that I am contending. How far I am from under-valuing the benefit to be expected from railways in their legitimate application will appear from the following lines published in 1837, and composed some years earlier.

Wordsworth (1906, p. 165) goes on to quote from his sonnet *Steamboats and Railways* of 1833:

Nature doth embrace
Her lawful offspring in man's Art; and Time,
Pleased with you triumphs o'er his brother Space,
Accepts from your bold hand the proffered crown
Of hope, and welcomes you with cheer sublime.

A sonnet is traditionally a love poem, but here the beloved is a machine. The pastoro-technical idyll of the Romantics is thus not all that far from the techno-pastoral ideal of the Futurists who wrote love poems to the car. In this sonnet, and in conformity to its conventional rhyme scheme, time and sublime rhyme, space and sublime do not. Time and the sublime also rhyme in the couplet motto of 'Is it not a feat sublime?/Intellect hath conquered time', part of the engraved masthead for *The Telegrapher*, the monthly review of the United States National Telegraphic Union first published in 1864 (Thompson, 1947, pp. 390–1).

The sublime transcends space and compresses time by speeding up travel by going faster through space in a shorter time. The railway entails 'the conquest of time and space' (O'Dell and Richards, 1971, p. 17). Railway culture, for Novak (1980, p. 166), was 'possibly the first modern culture to fulfil the ancient desire to compress space and purchase time'. The United States was the torch-bearer for this desire and its national embodiment. In the Unites States, for Martin (1992, p. 31), 'the railroad's awesome potential for minimizing the constraints of space and time were exploited to the fullest' in what Nye (1994, p. 57) calls 'the dynamic technological sublime [which] focused on the triumph of machines – particularly the railroad, but also the telegraph – over space and time'.

The train rocks to the rhythm of the magical machine travelling through space, but the sublime rhymes with time and transcends space. With the railway, as Alfred de Vigny put it, 'Distance and time are now conquered by Science, / Which encircles the world with its road sad and straight' (cited in Benjamin, 1999a, p. 716). As every walker knows, there is no road sadder than a straight road; a happy road is a winding road. 'The long and winding road that leads to your door', as Lennon and McCartney put it, may be a poignantly melancholic and love-lorn road, but it is better than the boring long and straight road that leads away *from* your door.

The railway also produced a new logistics of perception that paved the way, or perhaps more precisely laid the line, for photography and the later vectors of communication to travel on. Photography, Danly (1988, p. 31) argues, was

> The principal medium through which the aesthetics of travel developed, both as advertisements and as travellers' personal mementos...Train travel itself becomes an aesthetic experience, a search for picturesque views to record with the camera.

The railway journey carried the cultural baggage of the European land-scape aesthetic.

In particular, the railway created a static view for the mobile gaze, but otherwise immobile body, of the seated passenger and thereby, as Nye puts it (1994, p. 53),

> changed the appearance of the local landscape. The slow unwinding view seen from a wagon or a horse was transformed into a sliding world that seemed to move by while the passenger sat immobile. The

eye was not prepared to see these hurtling objects glimpsed in a rush, and had to learn to focus on the distant panorama.

Indeed, the railway constructed the distant landscape as panorama. And the panorama constructed the railway landscape as panoramic.

The railway created a new landscape by stage-managing the land into a panoramic landscape (see Kasson, 1976/1999, p. 116; Schivelbusch, 1986, pp. 60–1). The railway was not only depicted in a panoramic landscape as in Cropsey's painting, but also created a panoramic landscape. The viewer of Cropsey's painting is positioned as if she or he were in another train running on parallel tracks to those of the train in his painting and viewing his landscape. Unlike the railway and the telegraph lines depicted in the painting which will converge at the same point in time and space outside the painting, the railway track depicted in the painting and the one on which the viewer is travelling will never meet. If Cropsey's painting is viewed in an art gallery, the viewer moves on to view other panoramic landscapes on parallel tracks to them. The gallery becomes diorama in which the scene stays still, but the viewer moves. The landscape is mediated by the distance between the viewer and the landscape (painting) and by the orderly succession of paintings through which the viewer passes.

The projectile of the train not only hurtled the passenger through the barrel of the train line but also produced the landscape as panorama viewed through the windows of transportation and communication technologies. The train window, for Foster (2003, p. 155), 'framed the national terrain primarily as something "seen," ... as *landscape*. Removing the individual from direct bodily engagement with it, the railway turned the terrain into something distanced and panoramic.' Sternberger (cited by Schivelbusch, 1986, p. 61) argues that 'the views from the windows of Europe have entirely lost their dimension of depth and have become ... merely a painted surface'. The windows of Europe in the form of the railway and the car have been exported to the rest of the world. They have also created the form of panoramic perception in which the world is now seen, including the window of the cinema and television screen and of the computer monitor and its software. The railway, for Schivelbusch (1986, p. 62), 'first and foremost, is the main cause for such [a] panoramization of the world'.

The railway is both projectile and panorama which separates the traveller/viewer from the place through which he/she is travelling and which he/she is seeing. Perception, for Schivelbusch (1986, p. 64), 'no longer belonged to the same space as the perceived objects' as panoramic perception produced both a disjunction between perceptual position and perceived object and what he calls 'a new, evanescent reality'. Similarly, for de Certeau (1984, p. 112),

> the iron rail whose straight line cuts through space ... transforms the serene identities of the soil into the speed with which they slip away into the distance. The windowpane is what allows us to *see*, and the rail, what allows us to *move* through. These are two complementary modes of separation. The first creates the spectator's distance. You shall not touch; the more you see, the less you hold – a dispossession of the hand in favour of a greater trajectory for the eye. The second inscribes, indefinitely the injunction to pass on ... – an imperative of separation which obliges one to pay for an abstract ocular domination of space by leaving behind any proper place.

As we shall see with photography, Benjamin bemoaned a similar and subsequently additional dispossession of the hand in favour of a greater trajectory for the eye.

By conquering distance and time, the railroad, as Chandler (1977, pp. 83 and 86) puts it, 'revolutionized the speed of travel ... for the first time in history, freight and passengers could be carried overland at a speed faster than that of a horse'. The railway simultaneously speeded up transportation and communication and so transcended the constraints of terrestrial space and local time. It made possible the faster production and reception of messages in two different places, faster than the fastest trained animal (horses or pigeons). As it speeded up communication and relied on a direct physical connection between two places, the railway also made possible the development of telegraphy.

The railway was, Walt Whitman (1945, p. 583) claimed in 1880, 'the conqueror of crude nature', or more precisely, it was the crude conqueror of nature. Yet the railway not only conquered nature but also colonised it. Railways 'through the bush', in the terms of Franz Fanon (1967, p. 201), the pioneer theorist of decolonisation and a leading activist for it, were both the coloniser of nature and 'the natives'. Colonisation of nature and 'the natives' for him were 'one and the same thing'. Colonisation is just as much about colonisation of nature as it is about colonisation of the 'natives' (see Giblett, 1996, chapter 3). Railways played a decisive role in both. Henry Lawson (1976, p. 78) described in the 1890s how 'The mighty bush with iron rails/Is tethered to the world.' 'Tethered' connotes tamed and domesticated, though it also implies the possibility of breaking loose and running free. The bush, its life and its energy, is enslaved, though not vanquished.

The railway used the revolutionary, modern industrial technology of the steam engine. This is a sublime technology as Henry David Thoreau (1997, p. 110) observed in 1854, in not so few words and not in strictly chemical terms. The steam engine transforms solid fuel (coal or wood) by boiling liquid (water) into gaseous steam pressure, the motive power (see O'Dell and Richards, 1971, pp. 78–95). Steam, for Marvin (1987, p. 56), was 'a powerful new energy source' which was 'set loose', in Henry Nash Smith's words (1950, p. 156), like a tethered animal by what he calls 'the technological revolution' in steam-powered transportation (see also p. 31). Railways let loose the tethered power of steam, but at the same time they tethered the bush to the wider world. The bush was tethered to the world by the railway not only because its commodities were transported by rail to distant markets but also because it provided the fuel to power the trains.

As trains go faster than humans and transport them and goods faster than other means of transportation (such as animals and barges), they were used in war very early on in their development, usually decisively by the side with the most rolling stock and the most extensive network of lines. The military use of railways began in the 1850s with the French in the Crimean War and then by both sides, though not equally, in the American Civil War in the 1860s (van Creveld, 1991, pp. 158–159 and 162). The railway, for McLuhan (1964, p. 102), 'raised the art of war to unheard-off intensity, making the American Civil War the first major conflict fought by rail'. In doing so, Martin (1992, p. 367) argues that the use of railways in this war 'proved that railroads did more to change the art of war than anything since gunpowder', and so were more revolutionary for him than the conoidal bullet. Railroads, Martin (1992, p. 369) concludes, played 'a major role in the creation of modern warfare'.

Railroads also played a crucial role in the winning of modern wars, beginning with the American Civil War in which they 'played a major part in the North's victory over the South' (Fishlock, 2004, p. 381). Martin goes on to relate that '*Victory rode the rails* (1953) is the title of an entire book on the railroads' role in the Civil War. The new mobility proved a remarkable advantage to those who first recognized and were best able to employ it.' The Union forces of the North did just that to their considerable advantage, and to the disadvantage of the Confederate forces of the South. Armies, as Cowan (1997, p. 117) put it, 'could travel faster to battlegrounds, as Confederate general discovered to their dismay'.

The railway increased the speed of transportation of messages, men and materiel in war and peace (see Fuller, 1943, pp. 4, 9 and 11; 1946/1998, pp. 116–117; 1961/1992, pp. 92–93; Martin, 1992, pp. 367–369). Stephenson, for Fuller (1961/1992, p. 93), 'more so than Napoleon or Clausewitz ['the father of modern war' as Fuller described him earlier (p. 12)] ... [,] was the father of the nation-in-arms'. Stephenson was thus a crucial figure in the beginnings of the dissolution of the separation between the military and the citizenry, the soldier and civilian, and the dominance of the former over the latter. The militarisation of civilian life characterises the uptake and use of transportation and communication technologies from the railway and the Civil War to the computer and the Cold War. Stephenson did so by encapsulating passengers in the projectile of the train and shooting them through the smooth barrel of the train line across the surface of the earth, staging it as panoramic landscape and reducing it to a theatre of military operations, of communications, logistics and speed thus destroying 'the serene identities of the soil', as de Certeau puts it.

Steam railways also allowed the unleashing of the forces that gave rise to the power of corporations, which one prescient contemporary critic of Stephenson recognised. For Henry George (cited by Smith, 1950, p. 191), writing in 1871, 'the gigantic corporations' are 'the alarming product of the new social forces which Watt and Stephenson introduced to the world'. These corporations, moreover, are being 'welded into still more titanic corporations'. Transnational corporations, the engines of globalization, have their beginnings in steam engines and steam railways.

The railway won 'the war against nature'. The crude conqueror of nature is a weapon of war. The inventor of the railway gave birth to total war. Stephenson's eldest child was the American Civil War, 'a war of astonishing modernity', for Fuller (1961/1992, p. 106), in which, amongst other things, 'the field telegraph' was tried out (as we will see in the next chapter) and 'armoured trains' were used. If 'movement is the soul of war' (as it was for Napoleon (see Fuller, 1961/1992, p. 114; see also p. 50)), then the railway was its body, its enactment, its vehicle and so it would become 'the most important factor in strategy' up to the First World War during which period it was seen as the means 'to conquer space by time' (Fuller, 1961/1992, pp. 114 and 118).

By transforming liquid and solid matter into steam power and by producing a new world in the process the railway was effectively a machine that gave birth. The locomotive was the brainchild of inventors and engineers. Along these lines, in 1884 Huysmans (1959, p. 37) asked rhetorically,

> Does there exist, anywhere on this earth, a being conceived in the joys of fornication and born in the throes of motherhood who is more dazzlingly, more outstandingly beautiful than the two locomotives recently put into service on the Northern Railway?

The birth of the locomotive from the brain of inventors and engineers was what has been called 'a Bachelor Birth from a Bachelor Machine'. Theweleit (1987, p. 330n; see also p. 315) describes Bachelor Births from Bachelor Machines as 'attempts to create a new reality by circumventing the female body – to engender the world from the brain'. The Bachelor Birth from a Bachelor Machine is a brainchild (see Sofia, 1992, p. 380). It entails what Theweleit (1989, p. 127; see also Giblett, 1996, Figure 1) later calls 'cerebral parthogenesis (the masculine form of the virgin birth)'. The birth of the locomotive was a technological analogue (technologue) of birth that appropriated the metaphors of procreation and parturition without acknowledgement of, let alone gratitude for, the generosity of mother earth for providing the materials to make the transformation of matter into power and the creation of a new world possible.

The first brainchild, and the first Bachelor Birth from a Bachelor Machine, was the birth of Pallas Athene, when she 'sprang out of Zeus' head as a full-grown woman' (Rose, 1928, pp. 50 and 108; Bolen, 1984, p. 76; Theweleit, 1987, p. 315). Yet Zeus swallowed Athena's mother Metis (wisdom, good counsel) when she was pregnant and then gave birth to Athena from his head. In classical patriarchal mythology Athena (or Pallas Athene) is the Goddess of defensive war (see Bullfinch, 1993, p. 132). Bachelor Machines not only give birth without the mother and her body, and without the body of mother earth, but also appropriate and consume the good things she produces just as Zeus tricked Athena's pregnant mother Metis (wisdom) into becoming smaller and swallowing her. Zeus incorporated Metis and 'took over her attributes as her own' (Bolen, 1984, p. 76), not only her wisdom but also her ability to give birth.

As with Pallas Athene, the brainchild is invariably a daughter, a 'daddy's girl', as she was with Huysmans' (1959, p. 37) locomotives on the Northern Railway:

> One of these, bearing the name of Crampton, is an adorable blond with a shrill voice, a long slender body imprisoned in a shiny brass

corset, and supple catlike movements; a smart golden blonde whose extraordinary grace can be quite terrifying when she stiffens her muscles of steel, sends the sweat pouring down her steaming flanks, sets her elegant wheels spinning in their wide circles, and hurtles away, full of life, at the head of an express or a boat-train. The other, Engerth, by name, is a strapping saturnine brunette given to uttering raucous, guttural cries, with a thickset figure encased in armour-plating of cast-iron; a monstrous creature with her dishevelled mane of black smoke and her six wheels coupled together low down, she gives an indication of her fantastic strength when, with an effort that shakes the very earth, she slowly and deliberately drags along her heavy train of goods-wagon.

Huysmans/des Esseintes cannot seem to make up his mind whether to figure the locomotives as either women or horses and ends up figuring them ambiguously as both, as feminised and fantasised iron horses split between the ethereal and sublimated blonde, the stereotypical star of the silver screen, and the gutsy brunette with grunt prepared to get down and get dirty.

The sublime is always shadowed by its secret or 'other', what Sofoulis sees as slime which she encapsulates in the parenthetical portmanteau 's(ub)lime' (see Giblett, 1996, chapter 2). The sublime train has its other in the miasmatic trains of Nabokov's *Lolita* that would 'cry in the monstrously hot and humid night with heartrending and ominous plangency, mingling power and hysteria in one desperate scream' (cited in Nicholl, 1997, p. 44). Trains mingle phallic power and feminised hysteria in one scream of mingled terror and horror, the ruling passions of modernity, mingling the terror of the Law of the Father and horror of the Body of the Mother (see Giblett, 1996, chapter 2). Trains are anthropomorphised, or at least animated, as criers in the monstrously hot and humid night that mingle the elements of fire and water, of heat and liquid, as the steam train itself applies fire to water to produce steam in the black mass of modernity in which the solid elements of earth, fire and water are transubstantiated (sublim(at)ed) into the gaseous element of air, and in which the element of fire, or electricity, is brought down to earth and harnessed in electrical telegraphy.

3
Mind Over Matter: Electrical Telegraphy

Jasper F. Cropsey's painting, *Sidney Plains with the Union of the Susquehanna and Unadilla Rivers*, not only depicts the pastoro-technical idyll of the railway and the landscape living together in organic harmony as we saw in the previous chapter. It also shows the railway line and telegraph line gradually converging to a point outside the pictorial frame as if they would inevitably come together at some other place in space and point in time in the future neither of which are part of this story, of this moment. Yet the railway and the telegraph had already converged as mutually beneficial technologies. Initially telegraph lines merely used railway right-of-ways and ran alongside railway lines (see Thompson, 1947, pp. 21 and 203; Martin, 1992, pp. 20–24).

Later telegraphy became a command, control and communication system for railways (see Marvin, 1987, p. 56; Morus, 1998, pp. 162, 195, 208 and 211). It took nearly a decade, though, from 1844 to 1852 for them, in Thompson's words (1947, pp. 22 and 203–216), to join forces to their mutual benefit and form what Martin (1992, pp. 23–24) calls 'the symbiotic relationship', or perhaps more precisely synergistic relationship, between them. Rather than convergence between communication technologies (such as between the computer and the telephone and telecommunications more generally) being a revolutionary innovation of the 1990s (see Gleick, 2002, pp. 3 and 29), it was very much the norm of the nineteenth century beginning with railways and telegraphy. This tradition was continued in the early twentieth century when television converged the broadcast medium of radio with the moving pictures of cinema to produce 'radio with pictures'.

Transcendental telegraphy

The electric telegraph extended and heightened the process of transcendence of local time and space begun by the railway. The railway
revolutionized the speed of transportation: for the first time in history
freight and passengers could be carried overland at a speed faster than
that of a horse. The telegraph revolutionized the speed of communication: for the first time in history messages could 'travel faster than a
messenger' (McLuhan, 1964, p. 89) and so faster than the carrier of the
messenger carrying the message (see Crowley and Heyer, 1991, p. 124).
Message and messenger were separated; transportation and communication diverged. Messages could be carried over land and sea over long
distances at a speed faster than that of a horse, a ship and a train, faster
than all forms of transportation (see Garratt, 1958, p. 644). Over shorter
distances, though, such as in telegraph offices themselves and within
cities, the pneumatic tube system was 'much faster than sending the
messages by telegraph' (Standage, 1998, p. 91), probably only because
of the time taken to hand-deliver telegrams from the telegraph office to
their destination.

The telegraph, Czitrom (1982, p. 11) concludes, 'split communication
(of information, thought) from transportation (of people, materials)'.
Telegraphy superseded the railway by separating communication from
transportation (Czitrom, 1982, p. 11; Carey, 1989, p. 203). It made
possible the production and reception of messages in two different places
at virtually the same time. This feature, and others, of telegraphy was
later to characterise radio and television. The telegraph, Carey (1989,
p. 204) argues, 'allowed symbols to move independently of and faster
than transportation'. The most popular symbols used in telegraphy were
those of Morse code in which each letter of the English alphabet was
assigned a dot, a dash or various combinations thereof. The telegraph
and Morse code superseded letter writing by hand and postal delivery by
horse or pigeon or railway as the fastest means of communication available as the speed of communication *by* letter was tied to, and dependent
on, the speed of transportation *of* the letter. The telegraph, as Carey
(1989, p. 204) puts it slightly differently, 'freed communication from the
constraints of geography', of terrestrial space. Telegraphy was extraterrestrial like railways. Yet whereas railways sublimated solid matter into
air and smoothed the surface of the earth so that the passenger had the
sensation of flying above the earth, telegraph wires carried information
above the earth and information flew around the world (eventually).

Telegraphy and railways came to be regarded as inseparable (Morus, 1996, p. 361) as telegraphy worked in conjunction with railways in railroad coordination, especially signalling (McLuhan, 1964, p. 250; Carey, 1989, pp. 201 and 203; Israel, 1992, p. 52). Railways depended on what Marvin (1987, p. 56) calls 'its utilization of a new form of information [in telegraphy] organized in an even newer energy form, electricity'. Steam railways and electrical telegraphy using old and new energy forms travelled side-by-side on parallel tracks and lines with the latter controlling the former. Electricity did not immediately supersede steam as an energy form but controlled it through communication. In the United States, Marvin (1987, p. 56) argues, the railroad 'could not have become a continental transport system...without national telegraphic coordination of safety and scheduling'.

Telegraphy was a command and control system not only for railways but also for manufacturing. Just as the railway helped to make the civilian into a soldier and militarise civilian life so the telegraph helped make the country into a factory and the citizenry into the proletariat. 'Just as the mechanical loom imposed uniformity of action in the cotton mill', Morus (1996, p. 365) argues, 'the telegraph would guarantee such uniformity on the railways'. The railway tethered the mighty bush to the world with iron rails, but it was a crude instrument that corralled nature. Telegraphy was the regulation mechanism for railways. Telegraphy was the cybernetic steersman that steered railways in the right direction, at the right time. Both were the major impetus for the development of standard time (see Kern, 1983, pp. 11–14).

Both also reduced the sense of distance between places. For Claude Chappe, the inventor of the optical or semaphore telegraph (the precursor of the electrical telegraph; see Flichy, 1995, pp. 7–28; Garratt, 1958, pp. 645–646; van Creveld, 1991, pp. 154–156), 'the telegraph shrinks distances and in a way joins an entire, huge population *into a single point*' (cited in Virilio, 1995, p. 40; see also Flichy, 1995, p. 9). Similarly railways also were, as Kirby (1997, p. 4) puts it, 'shrinking the time and distance between the point of production and the point of consumption'. As she argues later, 'the effect of the shrinkage of space was to convert space into time' (Kirby, 1997, pp. 48–50). Space became time, or more precisely, the distance between places was calculated in terms of the time required to travel, or communicate, between them; the distance between places was converted into the time to travel by train or to send a telegraph between them.

How long it takes to travel between places becomes more important than how far apart they are; time takes precedence over space. The construction of railroads and telegraph lines for Jenkins (1987, p. 21) 'suddenly and dramatically shrank the earth ... railroads destroyed the protective geographic barriers of local markets'. Telegraphy for Standage (1998, pp. 1 and 97) had the effect of 'shrinking the world faster and further than ever before'. The railroad and telegraph builders' proud boast, or shameful confession, to their wives after a hard day's work on the railway or telegraph line was, 'Honey, I shrank the earth!' With the invention of the car, the plane and the radio what Riordan and Hoddeson (1997, p. 27) call 'practical American know-how had begun to shrink the world', though the railway and telegraphy had already begun to shrink it.

By making transportation and communication reliable, railways and telegraphy were a part of what Chandler (1977) calls 'the managerial revolution', and the associated manufacturing and marketing revolutions. They provided what he goes on to describe as

> the fast, regular, and dependable transportation and communication so essential to the high-volume production and distribution – the hallmark of large modern manufacturing or marketing enterprises. As important, the rail and telegraph companies were themselves the first most business enterprises to appear in the United States.
>
> (Chandler, 1977, p. 79)

By speeding up communication and making it faster than transportation, telegraphy superseded railways as the fastest form of communication. Chandler (1977, p. 89) goes on to argue that 'where the railroad improved communication by speeding the movement of mail, the telegraph and then the telephone permitted even faster – indeed almost instantaneous – communication in nearly every part of the nation'.

Yet the association between railways and telegraphy was not accidental – it was an arranged marriage. According to Ray (online),

> from its very beginnings, the transmission of intelligence over wire was closely allied with the railroad rights-of-way. The early histories of the railroad and the telegraph in the United States are so intertwined that the story of one cannot be told properly without touching upon the other.

Indeed, the intertwining story has to be told from the very beginning. Ray goes on to describe how on May 24, 1844, the world's first telegraph message, 'What Hath God Wrought,' which was written by Anne Ellsworth, daughter of the Commissioner of Patents, was sent over the line by Morse. The next day, a telegram was sent from Washington to the Baltimore 'Patriot.' Thus, the commercial use of the telegraph began on the railroad right-of-way.

What any Christian reader of the newspaper knew was the second half of the statement: 'let no man put asunder'. This statement alluded to the Anglican marriage ceremony: 'what God has joined together, let no man put asunder'. The human invention of telegraphy was given divine legitimation and a divine interdiction was put on sundering telegraph lines. Human ingenuity and prowess arrogated to itself divine power and rhetoric just like railways with Forrest as we saw in the previous chapter. Telegraph lines alongside railway tracks also applied in Great Britain, where 'by 1848 about half of the country's railway tracks had telegraph wires running alongside them' (Standage, 1998, p. 60). Winston (1998, p. 23) concurs that 'the earliest telegraph wires did indeed run beside railway tracks and were used for operational purposes', in particular 'to help manage the railways' (van Creveld, 1991, p. 157).

As well as this spatial convergence between railways and telegraphy, they were divergent, not only in speed, and so temporally divergent, but also in medium. Railways used steam engines on tracks laid on the earth, whereas telegraphy used electrical signals transmitted along wires suspended on poles above the earth. The telegraph for Essig (2003, p. 15) 'carried messages at the blazing speed of electricity'. Livingston (1996, p. 6) defines electrical telegraphy as 'the form of technology whereby coded messages could be sent over great distances, using electrical signals transmitted along wires'. Morus (1998, p. 211) defines telegraphy in similar terms as 'a multipurpose device for the instantaneous communication of information at a distance'. Telegraphy, as Virilio (1994b, p. 6) points out, means literally '*to write at a distance*'. The English term 'telegraphy' derives from the French 'télégraphe', or literally 'far writer', coined by Miot de Mélito in 1791 (Standage, 1998, p. 10). Writing from afar both acknowledges and overcomes distance. Similarly photography, or 'light writing', acknowledges and eliminates distance, and telephony means the transmission of sound over distance (see MacKechnie Jarvis, 1958, p. 226). Telegraphy, photography and telephony all entailed the overcoming of distance, and the overpowering of place.

Electrical telegraphy initiated a communication revolution just as the steam railway introduced a transportation revolution. The telegraph was what Carey (1989, p. 202) calls 'the first electrical engineering technology'. For Marvin (1988, p. 3) it was 'the first of the electrical communication machines' that entailed 'as significant a break with the past as printing before it'. For Standage (1998, p. 2) too, 'the telegraph unleashed the greatest revolution in communications since the development of the printing press'. It is the archetypal modern electrical communication technology as it was the first to separate terrestrial space from local time. In its heyday it was called 'the most perfect invention of modern times' (cited in Moyal, 1984, p. 17). It was also, according to Standage (1998, p. 197), 'the first technology to be seized upon as a panacea'. By no means was it the last. Each succeeding communication technology has been seen as a panacea.

The heyday of electrical telegraphy was from the early 1840s to the 1920s following the taking out of the patent in 1837 (Livingston, 1996, p. 6; Morus, 1996, p. 339). Livingston (1996, p. 6) goes so far as to call this period 'the age of the telegraph'. The mid-nineteenth century was also the 'age of the railway' (p. 44). The age of telegraphy outlasted the age of the railway and coincided from 1880 on with 'the age of "new imperialism"... when the imperial European powers successfully scrambled for overseas territory in Africa, Asia and the Pacific' (Livingston, 1996, p. 9). Telegraphy was closely aligned with both railways and imperialism. Both telegraphy and railways were also closely aligned with war and were used extensively in war. Chappe saw telegraphs, 'before all else, as instruments of war' (Gerspach cited in Flichy, 1995, p. 9). The Crimean War (1854–1855) was 'the telegraph war' as it was, for Standage (1998, p. 146), 'the first in which the telegraph played a strategic role'.

Not only was the American Civil War (1861–1865) the first major conflict fought by rail as we have seen in the previous chapter but also, for Whitney-Smith (1996, p. 64), it was 'the most telling war of the early telegraph period'. The Civil War for Cowan (1997, p. 152) was 'the first in which military strategy depended on the quick flow of battle information over telegraph lines'. The disparity in the use of telegraphy by both sides was a decisive factor in the outcome as were railways as we saw in the previous chapter. Both sides developed a military telegraph system, but one side, the eventual winner (hardly surprisingly), was far more successful in doing this than the ultimate loser. By the end of the fiscal year 1862–1863 the Union forces of the North had over 5000 miles of telegraph line in operation. Over one million telegrams had been sent during that year along these lines at an average of over 3000 per day. Over the next 2 years of the war, and so by the end of the war, the length of the Union military telegraph system had increased to about 15,000 miles compared to the Confederate forces of the South's 1000 miles. The telegraph accelerated the mobility of the Union army, and assisted in the construction and operation of military railways, and so contributed to their victory (see Thompson, 1947, pp. 393–394; see also Israel, 1992, p. 55).

Both railways and telegraphy were also used during the Boer War (1899–1902) in South Africa, when and where, as Bailes (1980, p. 67) puts it,

> the railway network guided strategy to an extent unknown in Victorian campaigning... [whilst b]y the end of the century the tactical use of telegraphy was a practised art... [T]hroughout the campaign telegraphs were used on a vast scale; over thirteen million transactions were recorded on 27 000 miles of main line and cable.

Livingston concludes that telegraphy was 'the most [*sic*] dominant of the communication inventions of the nineteenth century'. Certainly it began early enough in the century and lasted long enough, outlasting the century, to justify this claim.

Telegraphy also had far-reaching effects on other communication technologies of the nineteenth century since, for Livingston (1996, p. 7), 'it led directly to three other great communication inventions in contemporary world history: Bell's telephone, Edison's phonograph and Marconi's wireless (radio)'. It also had ramifications into the twentieth century and its communication technologies. For Winston (1998, p. 25), the telegraph led to 'the fax machine and television'. It even led beyond to the computer. The telegraph, for Carey (1989, p. 202), prefigured the computer which, for Marvin (1988, p. 3), is 'no more than an instantaneous telegraph with a prodigious memory'. The digital computer and the Morse telegraph both use binary codes. The telegraph, Carey even suggested later in an interview (cited in Livingston, 1996, p. 7), was 'the first computer, with the Morse code "dot" and "dash" representing the first use of binary logic in a large-scale technological system'. Dots and dashes led to 1s and 0s (see Standage, 1998, p. 8). The telegraph, Winston (1998, p. 29) concludes, was 'the model of all the signalling systems which follow'.

Taking its progeny in both the nineteenth and the twentieth centuries into account may justify Porter's and Teich's (cited in Livingston, 1996, p. 7) claim that 'telegraphy is the primary technological innovation that made the information age possible'. Certainly, for Moyal (1984, p. 71), 'the telegraph was... the foundation of the new information society in

Australia'. No telegraphy and there would be no Internet, no computers, no telecommunication satellites, no television, no telephony, no radio. Only railways. And not very efficient or safe ones at that as telegraphy controlled and regulated them. The telegraph prefigured the Internet by being not only generally what Standage (1998) has called recently (with the benefit of hindsight) 'the Victorian Internet', but also specifically the means by which the United States was, what Thompson (1947, p. vii) called over 50 years ago (with prescient foresight), 'hastily webbed with a crude network of wires'. Before the Internet and the World Wide Web was the network and web of telegraph wires.

The perhaps unlikely prophet of the succession from railways through telegraphy to the Internet was Walt Whitman in 1871 (1945, pp. 275–279), who in his poem 'Passage to India' had no qualms about

Singing the strong light works of engineers,
Our modern wonders, (the antique ponderous Seven outvied,)
In the Old World the east the Suez Canal,
The New by its mighty railroad spann'd,
The sea inlaid with eloquent gentle wires...
The earth to be spann'd, connected by network,
...the distant brought near,
The lands to be welded together...
All these separations and gaps shall be taken up and hook'd and link'd
            together...

Modern transport and communication technologies reversed 'continental drift' (or what we today call 'plate tectonics') to create what Serres calls 'plaque tectonics' (Conley, 1997, p. 65; Serres, 1995b, p. 16; see also Giblett, 2004), in which great shifting forces of modern cities and their transportational and communication infrastructure cover the globe and permeate the electromagnetosphere and orbital extraterrestrial space. If the railway tethered the mighty bush to the world and was the conqueror of crude nature, the telegraph went one step further and exploited the magical powers of nature. The railway enslaved nature and telegraphy put the slave to work.

Telegraphy communicated either via singing wires strung across the surface of the earth or via 'magic' submarine cables (Moyal, 1984, p. 35) laid in the depths of the ocean. The surface of the earth and depths of the ocean were colonised by cable and wire. In 1840 the Bishop of Llandaff described how the telegraph far exceeds even the feats of pretended magic and the wildest fictions of the East. This subjugation of nature and conversion of her [*sic*] powers to the use and will of man [*sic*] actually do, as Lord Bacon predicted it would do, a thousand times more than what all the preternatural powers which men have dreamt of and wished to obtain were ever imagined capable of doing.

(cited in Morus, 1996, p. 340)

Telegraphy was the white magic and the tame fiction of the West that fulfilled the dreams of the men of science of conquering nature and using 'her' power to their advantage. Ashantee tribesmen of Africa reportedly regarded the telegraph as 'the white man's fetish' (Moyal, 1984, p. 54), an object that the white man both imbued with prestige and used to create magic by wielding power. Yet Christian leaders in Baltimore found the telegraph 'too much like black magic for their liking' (Standage, 1998, p. 53). Congressmen in Washington found it to be more 'like a conjuring trick than a means of communication' (Standage, 1998, p. 43), though magic is a means of communication with another, invisible world. Amerindigenes also expressed distrust and suspicion of 'the magic of the talking wires' (Brown, 1970, p. 126), which made them strange bedfellows with those Christian leaders and Congressmen who expressed similar sentiments, and which was probably one of the few things they were ever likely to agree upon. Such sentiments persisted and were reproduced elsewhere when an historian of the Crimean War referred to the telegraph as 'that new and dangerous magic' (cited in Standage, 1998, p. 146).

Electrical sublime

Electricity was the means to create and wield this magic. Electricity, for Mosco (2004, p. 123), 'became the white magic of science, helping to win the battle against both the darkness of night and the evil darkness of black magic'. The electrical sublime created what Nye (1994, pp. 152–153) calls 'a synthetic environment infused with mystery... [T]he electrical sublime produced awe on demand.' Electricity was both part of the process of sublimation and the product of sublimation. Electricity, for Essig (2003, p. 28), enabled 'the transformation of energy from black coal to white light'. It was the fire and heat that enabled sublimation to take place. Discussing what he calls 'electricity's erotic quality', Metzger (1996, p. 16) argues that, 'as a love-toy, a weapon and a force that could perhaps instil life in non-living matter, electricity was conceived as an elixir – mysterious, powerful, unpredictable'.

In this modern mythology of technology Edison or Morse was the modern Prometheus who 'drew [if not stole like his Titanic predecessor] from heaven the strange, fierce fire' (cited in Standage, 1998, p. 23) and brought it down to earth, where he placed it in human hands, if not hearts, and later in bodies in the cinema as we will see in a later chapter. According to one of Morse's contemporaries, 'the electric spark is the true Promethean fire which is to kindle human hearts' (cited in Standage, 1998, p. 98). And according to one of Edison's contemporaries, he 'tamed the thunderbolt' (cited in Essig, 2003, p. 206). A century and a half later Martin (1992, p. 31) invokes the same mythology when he claims that 'modern man's grasp of electricity almost makes Prometheus' firebrand look antique by comparison', and indeed it is, though it is arguably the same substance, or electricity is the modern equivalent.

To another of Morse's contemporaries, 'the electric wires...web the world in a network of throbbing life' (cited in Standage, 1998, p. 159). Electricity is fire in the high modern lexicon of the elements; electricity is the divine spark of life in the secular theology of modernity (or what Carey with Quirk (1989a, pp. 114 and 115) calls the 'secular religiosity' of 'secular theologians'). Electricity is the unholy spirit that breathes modern life, or at least warmth, into otherwise cold and dead human hearts; electricity is the god that creates life, or at least makes it throb; electricity is the fire in the belly that drives the monster of modernity. No longer would we have earth, air, fire and water, the traditional four elements of Western philosophy going back to the pre-Socratics, but land, gas, electricity and liquid, the new elements, and commodities, of modernity. The elements were abstracted into material qualities and secular forces to be commodified by being industrialised.

Telegraphy, for Marvin (1988, p. 9), was part of an 'industrial shift from steam to electricity taking place in the United States and Western Europe at the end of the nineteenth century'. Electrical power, for Hanson (1982, p. xii), was also 'the means by which the United States shifted from an agricultural to an industrial society' in the late nineteenth century. Yet electricity was not only an industrial source of power but also what Marvin calls 'the transformative agent of social possibility'. Electricity was metaphysical magic. It had what she goes on to call 'the vitality of a natural force' (Marvin,1988, p. 63). Yet for some electricity was ambiguous. For the Reverend Ezra Gannett (cited in Czitrom, 1982, p. 9; Carey, 1989, p. 206), it was both the 'swift winged messenger of destruction' and the 'vital energy of material creation'.

Electricity encapsulates the ambiguity of modernity, and modern industrial (including communication) technologies: it is both destructive and creative.

Telegraphy was the means of harnessing and using the creative and destructive force of electricity. It did so, Thompson (1947, p. 3) claims, 'during the brief span of half a century' in which 'man' was 'to accomplish more towards harnessing the forces of nature than in similar period since the birth of civilisation'. Electricity was a force of nature and telegraphy the means to harness it. Telegraphy was the first technology to use what Crowley and Heyer (1991, p. 124) call 'harnessable electricity'. Electricity was a wild force like a horse that had to be tamed and harnessed in order to be exploited. As the railway tamed and tethered the bush to the world so the telegraph harnessed electricity for the world of business and government. Both sublimated natural animal power into technology.

Telegraphy was bruited in what Carey and Quirk (1989a, pp. 123 and 206) have called the 'rhetoric of the electrical sublime'. The Reverend Mr Williamson was not only concerned with the power of the mechanic arts to smooth out terrestrial space as we saw in the previous chapter with the railway, but also praised the power of electricity and the use of the electromagnetosphere to triumph over space. In 1841 he saw that the mechanic arts 'threaten to seize upon the forked lightning, and pluck from the faithful magnet a power that shall... achieve a yet mightier triumph [than sailing ships had] over the obstacles that space has interposed to the intercourse of man with his fellow-man' (cited in Marx, L., 1988a, p. 133). Telegraph lines were described in the nineteenth century in the United States as 'lightning lines' (Thompson, 1947, p. 204) and telegraphy itself as 'writing with lightning' (Martin, 1992, p. 20). One journalist claimed that 'we have trained the electric agent as a dutiful child or obedient servant, to carry our messages through the air by the road we have made for it' (cited in Morus, 1996, p. 341; 1998, p. 194). Another journalist proclaimed that what 'had been a scarecrow and chimera, began to be treated as a confidential servant' (cited in Standage, 1998, p. 59).

The railway cut through stone to make a way of transportation through the air, tethered the bush to the world and domesticated and tamed a wild beast. The telegraph, by contrast, cut across the surface of the land to make a way of communication through the air, chained the colony to the imperial centre and milked the domesticated animal. It was not only, as Morus (1996, p. 341) argues, 'unruly capriciousness tamed', but also the energy of electricity enslaved and trained to be a dutiful daughter to carry messages through space. Nature and the earth were infantilised as a recalcitrant girl who needed to be disciplined. Her unruly body needed to be contained by telegraphy, which for both past and recent cartoonists and commentators 'put a girdle round about the earth' (cited in Morus, 1996, p. 341; reproduced in Barty-King, 1979, p. 24; see also Moyal, 1984, p. 35).

Telegraphy was even for one poet (cited in Standage, 1998, p. 79) 'a loving girdle round the earth'. Tributes to Morse claimed that telegraph wires were 'girdling the earth' (cited in Standage, 1998, p. 170). The Governor of New South Wales inaugurating the overseas connection with Sydney in 1872 went further to say that 'the earth has been girdled with a magic chain' (cited in Moyal, 1984, p. 56). Shakespeare may have been responsible for this rhetorical flourish as the first issue of *The Telegrapher* published in 1864 quoted him as boasting that 'I'll put a girdle around about the earth in forty minutes' (Thompson, 1947, p. 399).

The overland extraterrestrial telegraph and the undersea submarine cable not only contained but also imprisoned the earth. Only those with the technical knowledge of the (Morse) code could unlock the chain to reveal almost all her secrets. Telegraphy was girdle, chastity belt and g-string. Like the railway, the telegraph as a girdle smoothed out the humps and hollows of the earth. It made the earth conform to the patriarchal and filiarchal ideal of youthful feminine beauty. It transformed the swampy depths of the old and implacable Great Goddess into the svelte surface of the young and benign virginal goddess of information and the star (see Giblett, 1996, especially chapter 2). It also enchained that body in a code that only the initiated could use to unlock and access her secrets.

The telegraph disciplined and constrained the feminised body of the earth in the realm of electricity and communication as drains were doing in the realm of agriculture and private property (see Giblett, 1996, especially chapter 1). Discipline and enchain supplemented and reinforced discipline and drain. Nature became the object of Victorian pedagogy. The electric telegraph, Morus (1996, p. 339) argues, was 'the perfect example of man's [sic] capacity to put nature's service at his disposal and reap the social rewards of that power' just as the earlier technology of ditching and draining had made 'nature move to an arranged design', as Williams (1973, p. 124) put it. The electric telegraph also, as Morus (1996, pp. 340–341) puts it, 'bound the empire together' and 'provided a technology that disciplined nature. As such it had the capacity to discipline society as well.'

In the United States the 'talking wires' (Brown, 1970, p. 68) of the telegraph penetrated the frontier and colonised the 'wilderness' beyond of Amerindigenes' homelands. In Australia and elsewhere the 'singing wire' (Carey, 1989, p. 202) or 'singing line' (see Thomson, 1999) of the telegraph not only colonised the songlines of Aboriginal country but also transformed the far-flung colonies into a cohesive empire. The railway tethered the bush to the world with its iron rails, but the telegraph bound the bush to the rest of the empire with its magic chains, talking wires and singing lines. The electric telegraph for two of its nineteenth-century historians even gave the body of the earth palpitations, hardly a healthy state of affairs, even though it might give it a kind of galvanic, Franken-steinian, monstrous half-life, when they proclaimed, 'the whole earth will be belted with electric current, palpitating with human thoughts and emotions' (cited in Barty-King, 1979, p. 13 and in Standage, 1998, p. 80). The possibilities of breaking free from the spell woven by the magic, of liberating enslaved nature, of returning the earth to its natural, unstressed rhythms, were reduced.

The cultural implications of disrupting the temporality of local place and terrestrial space were profound. Summing up the major, recent developments in transportation and communication, Théophile Gautier proclaimed in 1858 that

> Space and time have ceased to exist. The propeller creates its vibrating spiral, the paddle-wheel beats the waves, the locomotive pants and grinds in a whirlwind of speed; conversations take place between one shore of the ocean and the other; the electric fluid has taken to carrying the mail; the power of the thunderstorm sends letters coursing along a wire. The sun is a draughtsman who depicts land-scapes, human types, monuments; the daguerreotype opens its brass-lidded eye of glass, and a view, a ruin, a group of people, is captured in an instant.
>
> (cited in Schwartz and Ryan, 2003, p. 2)

Similar sentiments were expressed by his contemporaries regarding tele-graphy, as we will see later in this chapter, and photography, as we will see in a later chapter. A poem written in tribute to Morse, 'the father of telegraphy' (cited in Thomson, 1999, pp. 208 and 233; see also Thompson, 1947, p. 5), suggested that

> Science proclaimed from shore to shore,
> That Time and Space ruled man no more.

Morse was eulogised towards the end of his life for having 'annihilated both space and time in the transmission of intelligence. The breadth of the Atlantic is as nothing' (Standage, 1998, p. 87; see also p. 147; Czitrom, 1982, p. 11; Carey, 1989, p. 207; Livingston, 1996, p. 46; Miller, 1966, p. 307; Morus, 1996, p. 341). The ability of the telegraph to make 'men thousands of miles apart privy to the same intelligence at almost the same time' is, for Martin (1992, p. 31), 'almost miraculous'.

The 'annihilation of space and time' was borrowed from one of Alexander Pope's relatively obscure poems. It is a stock phrase that appears more often than any other in what Leo Marx (1964, p. 194) calls 'the entire lexicon of progress' as 'the extravagance of this sentiment apparently is felt to match the sublimity of technological progress.' Yet the phrase not only matches rhetorically the sublimity of technological progress; it also describes it physically, and chemically. Annihilating anything, making something into nothing, is a process of sublimation counter to the process of creation, creating something out of nothing (*ex nihilo*). By creating sublime railways and telegraphs 'men' were not being the divine creator who created something out of nothing. They were creating something out of something else, transforming existing matter into other matter. By sublimating solid matter into gaseous steam and electrical impulses, transforming something into airy nothing, they were in fact countering the divine fiat and creation *ex nihilo*. God does not sublimate, only men do.

The 'sublimity of technological progress' is a euphemism for what Karl Marx in the *Grundrisse* (the economic manuscripts of 1857–1858) saw as the striving of capital to annihilate 'space by time'. For him this stock phrase was no mere stylistic flourish but a powerful force of capitalism. In general, for Marx (1986, p. 109), 'ultimately all economy is a matter of economy of time' that regulates space, whereas specifically capitalist economy strives to annihilate space by time:

> The more production comes to be based on exchange value [and so less on use value], and thus on exchange [and so less on use], the more important for production do the physical conditions of exchange become – the means of communication and transportation. By its very nature, capital strives to go beyond every spatial limitation...: space must be annihilated by time.
>
> (Marx, 1986, p. 448)

Capitalism strives to annihilate space by time by the means of communication and transportation technologies. They are not ideologically neutral but are capitalist to the core.

By developing and using these means, capitalism destroys local economy and creates a world market as Marx went on to argue that

> while capital must strive on the one hand to tear down every local barrier to traffic, i.e. to exchange, and to conquer the whole world as its market, it strives on the other hand to annihilate space by means of time, i.e. to reduce to a minimum the time required for the movement [of products] from one place to another.
>
> (Marx, 1986, p. 463; see also Stratton, 1997, p. 254)

For one hundred years from the steam sublime of the railway (see Schivelbusch, 1986, pp. 33–44; Kirby, 1997, pp. 2, 44, 48 and 53) through the electrical sublime of the telegraph and radio to the electronic sublime of computers and the Internet, the rhetoricians of transportation and communication technologies have been in the business of proclaiming this annihilation (see Carey with Quirk, 1989a, pp. 115, 120, 134 and 139; Morus, 1998, p. 195). Of course, space and time will never be annihilated, but the triumph of time over space may have been completed. Indeed, Virilio (2000b, p. 47; see also pp. ix, viii and 3) argues that 'the depth of real time wins out over the depth of the real space of territories'. The capitalist modern sublime strives for the infinite spatiality of global time, or near instantaneity, or 'real-time', that overcomes the infinite temporality, or eternity, of local time.

The same rhetoric and practice reached my own far-flung corner of the empire when

> a 12-mile wire was hoisted [from Perth] to Fremantle and opened on 21st June 1869...15 years after the first line in Australia was opened in Victoria. First Telegram transmitted: "to the chairman of the Fremantle town trust. His Excellency Colonel Bruce Heartily congratulates the inhabitants of Fremantle on the annihilation of distance between the port and the capital and he requests that this the first message may be publicly known. Government house 21st June 1869."
>
> (Larry Rice http://www.omen.com.au/~larry/mfwahis1.html)

Of course, time and space had not been annihilated nor could 'the telegraph...destroy distance' (Standage, 1998, p. 101) literally, but telegraph unlocked the coupling and coordination of local space and time and reduced the sense of distance between places by reducing drastically the time taken to communicate between them. Radio, television, telephones, satellites and the Internet have only extended and heightened this process.

By freeing communication from the constraints of geography, telegraphy was part of the technological sublime that character-ises modernity (see Giblett, 1996, especially chapter 2). The sublime (including the sublime communication technologies of railways and telegraphy) transcends local place and physical body. The sublime can be defined simply as 'the exalted (originally what is carried aloft)' (Benjamin, 1999a, p. 415). The sublime means literally to carry the exalted aloft *under* the threshold, hence its relationship with the sublim-inal. The sublime and subliminal can be contrasted with the *superlim-inare*. Buber (1970, p. 173) defines the *superliminare* as the lintel *above* the door, the 'over-threshold', figuratively 'the sphere of the spirit'. The sublime, by contrast, is the sphere of the secular sacred, of 'the fetishism of commodities', as Marx put it, and of 'the phoney spell of the commodity', as Benjamin called it.

Carrying information aloft does not mean carrying it at a great height, though when it comes to communication satellites in orbital, extrater-restrial space, the height is quite considerable. Sublime communication technologies carry information aloft. In the beginnings of these tech-nologies information was carried aloft on railway and telegraph lines, but not especially high. Telegraph lines in war could easily be brought down so that 'they no longer ran aloft in the air'; railway lines could easily be blown up 'into the air' (Blunden, 1928, p. 25). The vulner-ability of both terrestrial means of communication and transportation in warfare was one of the drivers in the development of extraterrestrial means. Both paved the way though for this move 'off planet'. Railways filled in lowlands, levelled uplands and crossed gorges and rivers and so on to carry information over the surface of the earth. Telegraphy carried information aloft in lines strung between poles above the earth. Radio transmissions carried information further aloft by electromagnetic waves in the 'ether' or later spectrum. Satellites launched by rockets into orbital extraterrestrial space were later to carry information even further aloft. Communication technologies have progressively carried information further and further aloft.

The idea of carrying aloft was inscribed in the first electric telegraph line to cross Australia from south to north and to connect Australia with the rest of the world as it was called the 'overland telegraph line'. The major telegraph lines in nineteenth-century Australia not only went over land but also went across desert for here was a predominantly flat, horizontal landscape that lent itself to some extent to the placing of poles and stringing of wires across it.

The telegraph was not only couched or shrouded in what Leo Marx (1964, p. 195; my emphasis; see also pp. 197, 214, 217 and 230) calls the religio-philosophical '*rhetoric* of the technological sublime'. The telegraph *was* the technological sublime, especially the technological sublime in the horizontal, and of all horizontal space. The telegraph was an instrument of colonisation, colonisation of horizontal space, and transcendence of that space. The telegraph colonised hitherto uncolonised lands.

The desert, or at least the land believed to be desert – the last land to be colonised, the last stronghold of indigenous peoples, the unknown land – was the horizontal landscape par excellence through which the telegraph could be carried aloft. Larry Rice (http://www.omen.com.au/~larry/tgram.html) argues that

> With the completion of the famed Australian Overland Telegraph Line in 1872 between Adelaide and Darwin, the honour to send the first telegram naturally fell to Charles Todd (later Sir Charles) who conceived the idea, planned it and supervised the mammoth task. He telegraphed: 'We have this day, within two years, completed a line of communication two thousand miles long through the very centre of Australia, until a few years ago a *terra incognita* believed to be a desert.'

The electrical sublime presented what Carey (1989, p. 206) calls 'the mystery of the mind–body dualism [which] located vital energy in the realm of the mind, in the nonmaterial world'. By implication, vital energy was *not* located in the realm of the body, in the material world. As a result, the material world, the natural world, including the human body, was drained of vitality and became lifeless matter, becoming, as Karl Marx (1986, pp. 402–412) put it, 'man's' [*sic*] inorganic body (see also Giblett, forthcoming). 'Man's' body becomes inorganic, becomes lifeless. The telegraph, asserted the *New York Times* in 1858, was the foremost discovery that subjugated 'matter under the domain of mind' (cited in Czitrom, 1982, p. 10; Nye, 1994, p. 62). The telegraph exercised mind over matter and is the precursor of the computer.

Wordsworth wanted to have a bet both ways with railways as we have seen but he did not when it came to modern industry in general, for in *The Prelude* in horror he

recoiled
From showing as it is the monster birth
Engendered by these too industrious times.
Let few words paint it: 'tis a child, no child,
But a dwarf man; in knowledge, virtue, skill

(V, 1805–1806, ll291–295, 1972, p. 184).

Why? Because in his hands 'the ensigns of the empire ... he holds, / The globe and sceptre of his royalties'. These

Are telescopes, and crucibles, and maps.
Ships he can guide across the pathless sea,
And tell you all their cunning; he can read
The insides of the earth, and spell the stars

(V, 1805–1806, ll328–333, 1972, p. 186).

These 'ensigns of empire' were not only literal telescopes but also the figurative 'cheap telescope' of the war correspondent of the imperial wars of the late nineteenth century, who staged and mediated the spectacle of 'foreign wars' for newspaper readers at home (see Kaplan, 2000, pp. 290–294). The war correspondent was the hero of his own romantic adventure story of imperial ventures in exotic landscapes. War and the war correspondent, media and the military, reinforce and support each other. One cannot survive and function without the other. Although the media have played a critical role in reporting some wars (such as Vietnam) and been prevented from reporting others adequately (such as the 1991 Gulf War) (for a critical survey see Young and Jesser, 1997), the media need the spectacle of war and the military need a spectator of war.

The 'dear telescope' of the war correspondent in front of the newsreel or television camera staging and mediating the spectacle of 'foreign wars' of the twentieth and twenty-first centuries for the spectator in front of the television set is the direct descendant of his nineteenth-century forebears, as is the 'very dear telescope' of the 2003 Iraq War 'embedded' with Coalition forces, or 'in bed with the Marines and the Cavalry', with his satellite phone, antenna dish, laptop computer, digital camera, DC-to-AC adaptor and four-wheel drive producing the 'war porn' of live, close-up, obscene and compelling coverage of the 'Battle for Basra' and the 'Battle for Baghdad' in which the bloody reality of the carnage (including that of 'friendly fire') is lost in the welter of images, alliteration and euphemism.

The construction of railway and telegraph lines depended on the older chemical, mineral, industrial and visual technologies of smelting, wire- and lens-making, surveying and map-making. From these the modern 'man' gained his pre-eminence as emperor of the earth and from these he derived his position as royalty and his royalty payments. He can read the intraterrestrial and the extraterrestrial but he cannot read the terrestrial, so in Wordsworth's words,

> Meanwhile old grandame earth is grieved to find
> The plaything, which her love designed for him,
> Unthought of: in their woodland bed the flowers
> Weep and the river sides are all forlorn

> (V, 1805–1806, ll346–349, 1972, p. 186)

Not least because telegraph poles may line their banks as in Cropsey's painting. In its heyday telegraphy was described as 'the greatest instrument of power over earth which the ages of human history have revealed' (cited in Standage, 1998, pp. 174–175). And it was, thus far. More, much more, was to come. The 'ensigns of empire' of railway and telegraphy became ultimately the 'engines of empire' of the computer (see Shurkin, 1984, p. 37). In the meantime, the older mineral, industrial and visual technology of lens-making converged with the newer chemical and industrial technology of pictorial reproduction in photography and cinematography. The dwarf modern 'man' holds a camera in his hands in whose aperture he can read the insides of the earth and spell the stars, and then see them, or her, projected on a photographic plate or paper or on a cinema screen.

4
Shooting the Event: The Camera is a Gun, Photography is a Shot

The relationship between communication and writing that emerged in the first chapter re-appears in relation to photography because, as Raymond Williams (1974, p. 10) points out, the term means literally 'light-writing'. Although the inventors of the technology did not use the word 'photography', they conceived the idea in cognate terms. In 1826 Nicéphore Niépce coined 'the generic word... *héliographie* or sun-writing/drawing' (Batchen, 1997, p. 63; see also Virilio, 1994b, p. 19). If this, or cognate terms, did not come to mind for his contemporaries, then at least the idea came to them, or so they said later. In 1870 Hercules Florence claimed that 'in 1832... the idea of printing with sunlight came to me' (cited in Batchen, 1997, p. 44). The idea is not to have humans (or a human hand) and a technology writing (via a pen or other instrument or tool) but light itself so that the process of writing would be free of human interference, and the object being written would record itself as it were with verisimilitude. Before the automatic writing of the surrealists in the twentieth century, nineteenth-century photography aimed to make light-writing the hand or the instrument that inscribed automatically the lineaments of an object.

In this chapter, I trace critically the history of photography by focussing on the camera as a gun for shooting (and 'killing') events. When events were frozen in a frame on film they produced a spatial analogue of the event. Now that these minor events are digitalised on video disc the recorded event has a digitally encoded, almost spaceless manifestation. These events are bodily events, events, in the case of media events such as sporting events, of the human body. I place this history within the context of the drive for sublimation of the body manifested in the camera and in a number of other sublime communication technologies. Like the gun, I argue that these technologies are all in the business of recording or producing death. The camera sublimates the living organism of the body into the dead matter of the image. It transforms the depths of the living body into the dead surface of the image. The ultimate event is death.

The invention of photography had profound effects not only on what Virilio (1989b) calls the 'logistics of perception' but also on human physiology. 'For the first time in the process of pictorial reproduction', Benjamin (1973b, p. 221; 2002, p. 102; 2003, p. 253) argues, 'photography freed the hand of the most important artistic functions which henceforth devolved only upon the eye looking into the lens'. With photography, the eye became the dominant organ of pictorial production (rather than *re*production as that came later with lithography) over the hand. Moreover, the hand was reduced to a device for holding and aiming the camera, focussing the lens, adjusting shutter and film speeds, manipulating levers and ultimately clicking buttons. The hand became the servant of the eye and an appendage to the machine, and the human body became a kind of prosthetic vehicle for the camera that enabled it to get around and take photographs.

Photography gave the human eye greater power, but at the cost of the diminution of the power of the hand. Photography, for Benjamin (1973b, p. 226; 2002, p. 105; 2003, p. 256), was 'the first truly revolutionary means of reproduction', not only pictorially but also physiologically with the eye overthrowing and wresting power from the hand in pictorial reproduction. For Barthes (1972, p. 12; see also Giblett, 1985, p. 123), 'God and the emperor had the power of the hand, man has the gaze.' 'Man', that creature of secular humanism, has not only wrested power from God and the emperor, but also shifted the instrument of power from the hand to the eye. Both hand and eye (rather than other areas of the body or other senses) are instruments of sublimation. Sublimation, for Brown (1959, p. 291), is 'a displacement upward [from the anal and genital zones] into other organs (above all the hand and eye)'. 'Man' may shift power from hand to eye but 'he' is still operating in a sublimated realm that 'he' shares with, or more precisely, had usurped from, God.

The aim of photography was to print or paint with sunlight, but to paint or write what? This question had been answered before the word had been coined or the idea conceived. The object, or 'what', of photography had been identified decades before the process, or 'how', was formulated, and before the word, or its cognates, was coined. In 1799 Anthony Carlisle (cited in Batchen, 1997, pp. 30 and 112) described the experiments of Thomas Wedgwood to 'obtain and fix the shadow of objects by exposing the figures painted on glass, to fall upon a flat surface'. Photography reduces heights and depths to surfaces, converts the three dimensions of an object in space into virtually two. Photography not only records an object in a place in space but also captures and freezes that object in a moment of time. The conjunction of an appearance of an object in time and space constitutes an event. Photography records an event. I define event simply as a moment in the movement of bodies of matter in time and space (see Flew, 1983, p. 115). These bodies are either inanimate like light or animate like humans.

Even if the object is still, light moves through wave and particle motion to produce shadows. Wedgwood concluded that 'the new method of depicting by a camera, promises to be valuable for obtaining exact representations of fixed and still objects' (cited in Batchen, 1997, p. 112). Photography fixed a static object and a moment in time on a flat surface. Later photography, such as that of Marey and Muybridge, fixed the sequential movements of a dynamic object in moments of time on a flat surface. With Marey, Virilio (1991a, p. 18) argues, 'light is no longer the sun's "lighting up the stable masses of assembled volumes whose shadows are alone in movement"' but the means of tracking the dynamic movement of masses in space and time. The emphasis, though, in early photography fell not so much on fixing the object itself, but on fixing the shadow of an object. In 1830 William Henry Fox Talbot (cited in Batchen, 1997, p. 91) referred to photography as 'the art of fixing a shadow'. The object was incidental, or coincidental. It was just required to produce a shadow.

The art, and act, of fixing a shadow had a profound impact on perceptions of space and time. Photography was, for Talbot, 'a "space of a single minute" in which space *becomes* time, and time space' (cited in Batchen, 1997, p. 91). The appearance of an object in space becomes fixed in a moment of time, and a moment of time is fixed in the configuration and manifestation of an object in space. Time and space are collapsed together in the photographic event. 'The noblest function of photography', one booster claimed in 1864, was 'to remove from the paths of science ... the impediments of space and of time' (cited in Ryan, 1997, p. 21). With photography, seeing the world becomes, according to Virilio (1994b, p. 21), 'not only a matter of spatial distance but also of the *time-distance* to be eliminated'. For their contemporaries, railways annihilated space and telegraphy annihilated time, whereas photography for its contemporaries went one step further and produced in its early days, as Benjamin puts it, 'a strange weave of space and time', what he called 'aura' (see also Caygill, 1998, pp. 93–94 and 102–103; Wolin, 1982, pp. 187–190 and 237–238).

Aura

Considering photography as simply 'light-writing' is good etymology but problematic aetiology for it implies the question of what sort of writing? For Virilio (1989b, p. 81), 'photography, according to its inventor Nicéphore Niépce, was simply a method of engraving with light, where bodies inscribed their traces by virtue of their own luminosity'. Yet in inscribing its traces the engraved body was engaging in a double process split between the writing of inscription and the writing of the trace as we saw in Chapter 1 (see also Giblett, 1996, chapter 3). The body engraved in and by the photograph always leaves its traces, just as living for Benjamin is 'a leaving of traces' (Adorno and Benjamin, 1999, p. 104; see also Benjamin, 1973a, p. 169). Photography, for Benjamin (1973a, p. 48; 2003, p. 27), 'made it possible for the first time to preserve permanent and unmistakable traces of a human being', but he/she is not necessarily inscribed in the process. Photography is light-writing, but the writing of photography is split between writing as inscription and writing as trace. Benjamin (1999b, p. 512; see also 1979, p. 244) argues that 'the first people to be reproduced entered the visual space of photography with their innocence intact – or rather, without inscription'. In other words, they entered it, and their image was reproduced, or traced, with their aura.

Photography as trace and without inscription is auratic. Benjamin (1999b, p. 518; see also 1973b, pp. 222 and 224; 2002, pp. 103–105, especially n.5, p. 123; 2003, pp. 253–256) defines aura as 'a strange weave of space and time: the unique appearance or semblance of distance, no matter how close it may be'. To what does 'it' refer? Grammatically it refers to 'distance', but how can distance be close? Or faraway, for that matter? *That* is aura, the closeness or proximity of distance when distance is overcome but still maintained. Aura is the experience of an object looming up and receding at the same time. In collapsing time and space together, early photography produced an auratic 'presence in time and space' (Benjamin, 1973b, p. 222).

Aura is akin in this respect to Freud's concept of the uncanny, a strange, spectral presence/absence. Just as the uncanny is the return *to* the repressed content or portion of the unconscious (see Giblett, 1996, chapter 2), so the auratic is the return to the repressed of what Benjamin (1979, p. 243; 1999b, p. 512; 2002, p. 117; 2003, p. 266) called 'the optical unconscious'. And just as psychoanalysis brought the former to light, photography brings the latter to light – literally:

> For it is another nature that speaks to the camera than to the eye: other in the sense that a space informed by human consciousness gives way to a space informed by the unconscious...It is through photography that we first discover the existence of this optical unconscious, just as we discover the instinctual unconscious through psychoanalysis.
>
> (Benjamin, 1979, p. 243; 1999b, pp. 510–511)

Just as the uncanny is by definition invisible but is made sensible, especially through the sense of smell (see Giblett, 1996, chapter 2; 2004, chapter 3), so the auratic is the invisible made visible.

Aura is also similar to Freud's concept of symptom in which the surface of the body of the patient (including their behaviour and dress) bears and manifests the traces of their psychopathology (see Giblett, 1996, chapter 4). Aura operates in the circuit of the uncanny and symptomatic: simultaneously returning to the repressed to return the repressed to the surface of the body. Aura is the expression in old photographs of a profound and unique moment in time and a place in space; aura is more than a mere event, but the imbuing of an event with ritual significance that transcends time and space. Commenting on a *c.*1850 photograph of Schelling, Benjamin (1999b, p. 514) argues that 'the creases in people's clothes have an air of permanence'. Even the evanescent creases in their faces have a similar air of permanence. Aura is the play of permanence and evanescence. Benjamin (1973b, p. 228; 2002, p. 108; 2003, p. 258) suggests that 'for the last time the aura emanates from the early photographs in the fleeting expression of a human face'. The light-writing of photography kills the object in photographing it, but the photograph always bears the traces of the living body of the subject whether it be human, animal, vegetable or mineral, though it will not necessarily convey its aura.

Trace and aura are similar, but Benjamin (1994, p. 586; 2003, p. 106; Adorno and Benjamin, 1999, p. 290) asserts that 'the concept of the trace is defined and determined philosophically in opposition to the concept of aura'. Trace is a chemical residue indicating current absence and past presence, whereas aura is a physical quality symbolising present presence. Benjamin (1999a, p. 447) differentiates them in the following way:

the trace is appearance of nearness, however far removed the thing that left it behind may be. The aura is appearance of a distance, however close the thing that calls it forth. In the trace, we gain possession of the thing; in the aura it takes possession of us.

In the early photographs, the thing photographed takes possession of us – we do not take possession of it. We bask in its aura. The thing is separate and distanced from us. The aura, for Benjamin (1999a, p. 314), is 'the aura of distance opened up with the look that awakens in an object perceived'.

Aura is the looking back of the object of the camera (and its 'gaze' and the gaze of the photographer) to the viewer of the photograph:

> What was inevitably felt to be inhuman, one might even say deadly, in daguerreotypy was the (prolonged) looking into the camera, since the camera records our likeness without returning our gaze. But looking at someone carries the implicit expectation that our look will be returned by the object of our gaze. Where this expectation is met...there is an experience of the aura to the fullest extent...Experience of the aura thus rests on the transposition of a response common in human relationships to the relationship between the inanimate or natural objects and man [sic]. To perceive the aura of an object we look at means to invest it with the ability to look at us in return.
>
> (Benjamin, 1973a, pp. 147–148; 1973b, pp. 189–190; 2003, p. 338)

The living body's transposition of aura to the dead matter of the camera was possible in the age of long exposure times but declined with the advent of the snapshot.

Aura may wane as objects get nearer; trace will never disappear as objects get further away. Things may decreasingly take possession of us, but we will continue to take possession of them. And imbue them with significance. The decline of aura is concomitant with the incline of commodity fetishism. In the former things take possession of us whereas in the latter we take possession of things. In this process things will leave their traces (on us and of us on them for 'living means leaving traces' (Benjamin, 1973a, p. 169; see also Adorno and Benjamin, 1999, p. 104)) and decreasingly wrap us in their aura. In the later photographs, we take possession of the thing itself – it does not take possession of us. It leaves it traces but we do not bask in its aura. Aura is an energy- or force-field that all objects possess and that is acknowledged and reproduced in ritual.

Benjamin (1999b, pp. 327–328) distinguishes three aspects of genuine aura: 'First, genuine aura appears in all things, not just in certain kinds of things.' Aura is the sacral quality with which all objects (including everyday objects and subjects) are imbued in traditional or pre-modern cultures. A vestige of that quality lives on in modern cultures in the fetishism of commodities. 'Second, the aura undergoes changes.' Aura is not fixed or eternal, it can wax and wane. The auratic significance with which objects are imbued ebbs and flows without ever becoming totally bereft of it. 'Third, the characteristic feature of genuine aura is ornament, an ornamental halo, in which the object or being is enclosed as in a case.' The auratic object is framed or contained. It is marked off decoratively and spatially from other objects with which it shares sacrality on a continuum.

The fundamental distinction between the auratic and the inscriptive does not correlate to that between nature and culture, or orality and literacy, but between the cultures of first, or worked, nature and that of second, or worked-over nature (see Giblett, 2004, chapter 1). Aura is pervasive in the former, vestigial in the latter. For Benjamin (1999a, p. 362), 'the decline of the aura and the waning of the dream of a better nature...are one and the same'. Both go hand in hand with the commodification of nature, the photograph and photography. Nature is drained of sacral significance, and imbued with capitalist value, just as the aura of human subjects and other objects declines. And just as the photography of events rises. For Heidegger (cited in Giblett, 2004, p. 43), the fundamental event of the modern age is the conquest of the world as picture. Photography contributes not only to that conquest but also to the fundamental event of the hypermodern age: the conquest of the world as commodity.

Aura declines as images of objects are reproduced and as exposure times decrease 'from Niépce's 30 minutes in 1829 to roughly 20 seconds with Nadar [in] 1860' (Virilio, 1994b, p. 21). Aura in photographs is a function partly of long exposure time and slow lenses and shutter speeds, and partly of the subject's intact innocence. But the two go hand in hand: the snapshot produces a guilty subject who knows he/she is having his/her photograph taken and a commodified object of the photograph that is bought and can be resold. Capitalist commodities can be decorative, but mass reproduction does not (and cannot) mark them off spatially from other mass-produced objects. Benjamin (1999a, p. 337 and p. 343) argues that 'for the decline of the aura, one thing within the realm of mass production is of overriding importance: the massive reproduction of the image'.

The aura of a work of art, in Benjamin's (1973b, p. 223; 2002, p. 104; 2003, p. 254) words, 'withers in the age of mechanical reproduction'. Aura is living matter so it can wither, or 'shrivel', as Benjamin (1973b, p. 233) elsewhere puts it, like a plant. The photograph is commodified, sold and bought, like the book or painting, and so is dead matter. The photograph is a commodity, or dead matter, by virtue of light writing the outlines of object on a flat surface. Living matter sublimated into air is transformed into dead matter (see Giblett, 1996, figure 1). The new photograph is split between its status and function as inscription and as commodity. Photography is split between the old photograph and the aura it possesses, and the new photograph with its vestiges of aura and traces of the object.

Fire

Aura is a kind of divine fire, the living fiery breath of God, or the Holy Spirit, that 'He' breathed into everything in creation according to the biblical book of Genesis. Photography is an attempt not only to wrest fire from the gods, but also to reproduce it. Neale (1985, p. 25) points out that 'there existed a whole theology of light as the trace of the noumenal in the real, as the mark of spirit in matter'. Divine light was incarnated in the photographic body of the son/sun of God. Solid matter was sublimated into the sublime image in the *camera obscura* only to be incarnated and inscribed in the sublimate of the surface of the photograph. And later animated in cinematic film. Photography was the logical next step to the *camera obscura*, whose image was, as Neale (1985, p. 20) puts it, 'fleeting, intangible, evanescent'. Photography in its beginnings was concerned primarily with recording the event of the shadow rather than the outline of the object. It was more concerned with the index of the presence of the object than with the object itself.

By invoking the power of the sun in his concept of heliography, Batchen (1997, p. 63) comments that Niépce incorporated 'a key metaphor for God's divine power and benevolence'. By using the power and light of the sun, photography is a kind of secular theo-technology. It is also a kind of industrial alchemy. Photography transforms the solidity and base matter of the photographed object into the gold of inscriptions on the surfaces of the photographic plate, film, paper and disc. Photography is a sublime communication technology. Using light, the light of divine revelation, reason and rationalist technology, analogue photography transforms the base matter of solid objects via a slimy emulsion of chemicals into an evanescent visual image engraved or inscribed on the surface of the base matter of plate, film and paper. Digital photography transforms the base matter of solid objects via the writing of digital code into an evanescent visual image inscribed on the surface of the base matter of disc. Photography sublimates solid matter into the thin air of sun-filled space only to transform and commodify it into the sublimate of dead matter. Photography thereby achieves an almost complete circuit in what I have called elsewhere the 'pyschogeocorpography of modernity' (Giblett, 1996, Figure 1).

By using the power and light of the sun, photography was not so much light writing as nature writing. Nature is a desacralised or secularised divine force (see Giblett, 2004, chapters 1 and 2). The invention of photography stemmed in part from what Szarkozski (cited in Batchen, 1997, p. 18) calls the desire that 'it might be possible to "snatch from the very air a picture formed by the forces of nature"'. Photography stems in part from the Promethean desire to snatch fire from the gods in the heavens and bring it down to humans on earth. The photograph is basically the imprint or inscription of the shadow made by an object suspended in sunlit space or sunfilled air. It is evanescent and ethereal like a mist, an emanation of nature, or a ghost, the relic of a bygone living body.

For light, however, to inscribe an object in a photograph the guiding hand of 'man' is needed to hold the camera and take the shot – though he wanted to abstract himself from the process. By 1838, Batchen (1997, pp. 33 and 34) relates, 'Daguerre was able to announce the invention of a workable process that he modestly called daguerreotype: "it consists in the spontaneous reproduction of the images of nature received in the *camera obscura*"'. The daguerreotype represented the outcome of the desire to allow nature to reproduce itself spontaneously, like spontaneous combustion and generation, to reproduce images of itself without human intervention. Human beings and human bodies would be sublimated out of the process and into the process to become as evanescent and ethereal as they claimed or desired the process to be.

Despite this desire for abstraction, photography also enacted the desire to produce a machine that would reproduce images of nature, like a kind of mechanical God creating Adam in his own image. In 1838 Daguerre claimed that 'the DAGUERREOTYPE is not merely an instrument which serves to draw Nature; on the contrary it is a chemical and physical process which gives her the power to reproduce herself' (cited in Batchen, 1997, p. 66). Nature is constructed as incapable of reproducing 'herself', though 'she' had been doing it successfully for millennia before photography. Nature is figured as needing the helping hand of gynecological man, like some kind of artificial inseminator, to reproduce herself.

Photography is 'a Bachelor Machine for a Bachelor Birth' (see Giblett, 1996, p. 50, n.7) in which the photographer and the viewer give birth together to a new world. Along similar lines, Barthes (1981, p. 81) suggests that

A sort of umbilical cord links the body of the photographed thing to my gaze: light, though impalpable, is here a carnal medium, a skin I share with anyone who has been photographed.

The photographer gives birth to a new world out of his brain box and his Box Brownie, or out of his box of magic tricks (they all amount to the same thing), like some kind of gynecological magician who weaves a magic spell and wrests from nature her secrets using light, much like telegraphy did with electricity and radio was to do with electromagnetism. For Barthes, photography is *a sort of* umbilical cord that connects the objectified body in the photograph to the subject's gaze. Rather than giving birth to life, photography gives birth to death.

Although William Henry Fox Talbot referred in 1840 to photography as 'an act of "natural magic"' (cited in Batchen, 1997, p. 62), the magic was modern black magic woven by men ostensibly using nature. Batchen (1997, p. 92) argues that 'a number [of contemporary journalists] used the term *necromancy* (communication with the dead) to describe the actions of both Daguerre's and Talbot's processes'. Necromancy entailed not only communicating with the dead but also communing with death. Barthes (1981, p. 9) sees photography as 'the return of the dead' and Batchen (1997, p. 172) even sees it '*as* death' itself. Death, for Barthes (1981, p. 92), 'must be somewhere in a society'. He wonders whether it is

perhaps in this image which produces Death while trying to preserve life. Contemporary with the withdrawal of rites, Photography may correspond to the intrusion, in our modern society, of an asymbolic Death, outside of religion, outside of ritual, a kind of abrupt dive into literal Death. ... With the photograph, we enter into *flat Death*.

Rather than sexuality (the Victorian orthodoxy of repression), death is the repressed of modern European society and its settler diaspora that returns in the photograph flattened out and deprived of ritual and symbolic significance, stripped of aura.

The photograph is both the recorder and the bringer of death. The appropriately named Cadava (1992, pp. 90–91) describes how, as

> subjects of the photograph, seized by the camera, we are morti-
> fied by it...the photograph tell us we will die...It announces the
> death of the photographed...the photograph is a grave for the
> living dead...[it is] the tomb that writes...[it is] the allegory of our
> modernity...the uncanny tomb of our memory. Photography is a
> mode of bereavement...the return of the departed.

Photography is not so much life that writes with light but the tomb that writes the death of the photographed.

Photography is a tomb for the living dead in that the photograph kills. But the photograph also gives life, or at least a strange kind of afterlife for its subjects. The photograph tells us we will die, but it also tells us that we can have a life after death engraved or inscribed in its surface. Photography is thus both a tomb for the living dead and a womb for the dead living. Photography, for Barthes (1981, p. 82), 'has something to do with resurrection'. The photograph, or other representation produced by a camera, digital or analogue, does not merely kill the human body by reducing it to an object and fixing its lineaments but gives it a strange kind of afterlife as the dead living.

The dead living is ultimately the star (and the athlete and sportsperson insofar as he/she has become a star as we will see in the next chapter on cinema) – the star is always already a dead star however much he/she shines or however brightly or long he/she may shine. A dying star continues to burn brightly in our skies even though it died light years ago. A star (in two senses) is a heavenly body (in two senses) (see Dyer, 1987; 1998). A sportstar or screenstar *is* a heavenly body (and does not *have* a heavenly body) by virtue of the fact that their image has a life independent of theirs. Their image floats in a sublimated realm discon-nected from them as the term suggests.

Photography not only kills objects by fixing them but also kills the subject of the look, and the human eye, by replacing it with the camera which stands in the place of, and is thus literally a prosthesis for, the human eye. With the camera, Comolli (1980, p. 123) argues, 'the human eye loses its immemorial privilege; the mechanical eye of the photo-graphic machine now sees *in its place*...the photograph stands as at once the triumph and the grave of the eye'. The photograph stands as at once the monument to the power of the eye and the tomb of the eye as it stands also as at once the triumph and the grave of the object being photographed. The object triumphs over time and mortality to live on in the photograph only to be mortified by it. Cadava (1992, p. 89) concurs that 'the photographic image conjures up its death ... The home of the photographed is the cemetery.' This was literally the case with David Octavius Hill, who produced a famous series of photographs of people who made the cemetery their home. Photography found its true home, the home of the dead and the living, now living dead.

Gun

As photography kills, the camera is a weapon. The activities of 'loading' and 'aiming' a camera and then 'shooting' a photograph or an event on film are associated with war and hunting. In short, the camera is a gun. Etienne-Jules Marey's multiple-exposure 'photographic gun' of 1882 looked like a rifle, and was held and 'fired' like one (for an illustration see Neale, 1985, p. 35). For Virilio (1988, p. 190; see also 1991a, pp. 16–17), 'there is a very close similarity between ... the Gatling Gun, the photographic revolver of Janssen, and cameras. The Colt .45, the chronophotographic gun – all these things are tightly bound up with each other.' The modern camera, Sontag (1977, p. 14) argues, is 'trying to be a ray gun'. It not only records light, but also shoots events as if it were projecting rays at an object. Chronophotography, 'literally, the photography of time', as Kern (1983, p. 21) points out, or more extensively, the writing of light-time, shoots events in time and space.

The camera is a sublimated lethal weapon. The camera, in Sofia's (1992, p. 381) terms, is like 'the gun [which] may be read as a metonym of speedy and lethal penetration across terrain and through flesh, a tendency related to the military investment in technologies of speed', including technologies of communication, I would add, like photography and cinematography. The camera, for Sontag (1977, p. 14), is sold 'like a car, as a predatory weapon' though, of course, it does not kill like a car so she concludes that it is 'a sublimation of the gun' and 'to photograph someone is a sublimated murder'.

The gun and the camera (and the car) are devices for overcoming and eliminating distance. The gun is a device for eliminating the distance between a bullet and its target. The camera is a device for eliminating the distance between a recording surface and an event. The car is a device for eliminating the distance between the driver and his or her destination. For Robins and Webster (1999, pp. 240–259), 'the elimination of distance' brought about by 'distance-shrinking technologies', especially communication technologies, has worked towards what they call 'the neutralization of space' and the loss of 'presence at a distance' replaced by presence at hand.

The development of photography and weaponry changed the relation between the hand, the eye and the weapon. Seeing and killing are joined; the camera is a prosthesis for seeing and shooting events, and the rifle is a prosthesis for killing by seeing. The function of the weapon, for Virilio (2002a, pp. 53 and 112), is 'first of all the function of the eye: sighting'. With the camera and the rifle no longer do you look the enemy in the eye unlike fighting with a sword. You watch his or her movements and his or her weapon through your weapon. No longer do you hold your weapon, such as a sword, away from the line of sight and wield it as an extension of the hand, arm and spine. The line of sight and the surveying or positioning instrument had already been aligned with the technologies of colonialism, 'the ensigns of empire' as Wordsworth called them, such as the sextant, telescope and theodolite.

The rifle is a prosthesis not only for killing by seeing, but also for eating as it is a consumer of living matter. It transforms living matter into dead matter. The rifle, for McLuhan (1964, pp. 341 and 343), is 'an extension of the eye and the teeth'. It is another instance of the way in which for him 'weapons proper are extensions of hands, nails and teeth' that 'come into existence as tools needed for accelerating the processing of matter', including ingesting, digesting and excreting it. By shooting, the camera and the gun constitute their wielder as living on the logic that 'I kill therefore I am (living).' The gun and camera constitute the object of the shot as dead and the subject of the shooter as living. They mediate between them. They are communication technologies. In Metz's (1982, p. 50; see Laplanche and Pontalis, 1973, pp. 229–231 and 349–356) psychoanalytic terms, 'the camera is "trained" on the object like a fire-arm (= projection) and the object arrives to make an imprint, a trace, on the receptive surface of the film-strip (= introjection)'. The camera is a device for articulating projection and introjection so that projection outwards via a projectile is linked to introjection inwards via inscription or trace.

By shooting events or at targets the subject kills both animal and human bodies (= excorporation) by reducing them to a pile of gristle or a strip of images and consumes bodies (= incorporation) as either flesh or image. The photographing subject is situated at, and constituted as, the point of intersection of photographic projection and introjection, excorporation and incorporation. If the subject only projected and excorporated, there would be nothing to tell them they existed; if the subject only introjected and incorporated, they would be destroyed by what is outside them. They excommunicate in order not to be excommunicated. The choice for the subject is simple: excommunicate or be excommunicated. The subject requires projection and introjection, excorporation and incorporation, to be interpellated as subject. The gun and the camera provide the technological means to do both. But in order to do so, the gun and camera wielder has become an attenuated body reduced to an eye, and a vehicle to get them around. Rather than the camera or the gun being a prosthesis for the eye, the body of the wielder has become an ambulatory prosthesis for them, a means for them to get around.

The gun and camera make space and time into events, and thereby master both. They bring events into their purview, make them thereby into targets and constitute them as objects in the discourse of mastery. Without them and the other ensigns of empire, the wielder of gun or camera is truly lost in space. The gun and the camera, the sextant and the theodolite, survey and coordinate space to master it and make it traversable, like the map with its grid of longitude and latitude, and the grid-plan town (see Giblett, 1996, chapter 2). Photography, Friedberg (1993, p. 30) concludes, 'offered a mobilized gaze through a "virtual real," [and thereby] changed one's relation to bodily movements, to history, and to memory'. Moreover, I would add, to place, to other living beings, to nature. Photography mobilized the eye in the war against nature, but each shot of the camera is static, whereas cinematography is the mobile eye of the war against nature in which the shots are continuous like a machine-gun.

If, indeed, the camera is a gun and photography a shot with that gun as I have been arguing, it is hardly surprising then that photography played a role in war shortly after its development. The First World War (1914–1918), Virilio argues, was 'the first Total War', 'the first truly technical war in Europe' (Virilio and Lotringer, 1983, pp. 8–9) and 'the first total war of humanity against man' (Virilio, 2000b, p. 55). It was so partly because it was a watershed for both the devastating effects of war on human communication and the close connection between communication technologies and war in aerial photography. The First World War, for Sekula (1984, p. 34), was

the first occasion for the intensive use of aerial photography for 'intelligence' purposes... With airplane photography... two globalizing mediums, one of transportation and the other of communication, were united in the increasingly rationalized practice of warfare... A third medium of destruction, long-range artillery, was quickly added to this instrumental collage, making possible bombardment – was well as image recording – at a great distance.

Just as transportation and communication converged in railways and telegraphy, so they did even more seamlessly in planes and photography. Photography, plane and bombardment, all operating at a distance from their 'targets', made them into a quilt or collage of fields and towns one moment and bombed them into piles of rubble in the next in the landscape of warfare. 'The secrets of war are written in the air', as Virilio (1989b, p. 75) puts it, in two senses. These secrets are evanescent, insubstantial and transient, and they are only obtained by aerial observation and surveillance whether by plane or satellite. The light-writing of photography inscribes on film the ethereal secrets of war written in the air.

Yet, for Virilio (1989b, pp. 69–70), the First World War was an important watershed in the development of communications technologies, not so much because of their use, but in the type of guns used. He goes on to argue that

> if the First World War can be seen as the first mediated conflict in history, it is because rapid-firing guns largely replaced the plethora of individual weapons. Hand-to-hand fighting and physical confrontation were superseded by long-range butchery, in which the enemy was more or less invisible save for the flash and glow of his own guns. This explains the urgent need that developed for ever more accurate sighting, ever greater magnification, for *filming the war* and photographically reconstructing the battlefield; above all, it explains the newly dominant role of aerial observation in operational planning.

With the single-loading gun (and camera) the line of sight and the weapon were momentarily aligned for taking a shot. With the rapid-firing gun (and the self-loading camera), however, the line of sight and the weapon become aligned and stay aligned. It produces what Virilio (1989b, pp. 69 and 20) calls 'the deadly harmony ... between the functions of eye and weapons' to the point where '*the function of the weapon is the function of the eye*'. Having the enemy in the gun sight, even in the line of fire especially with the machine gun, means the soldier can (and will try to) kill him or her.

Not only is the camera a gun, but also the eye is a gun. Galvin (1994, p. 178) comments on Virilio's discussion that 'a line of sight ... is a line of fire. It is a perceptual trajectory along which information flows

in one direction [like a camera] and deadly force travels in the other.'
Virilio calls this two-way line of information and deadly force 'a vector'.
With the rapid-firing gun the line of sight and the weapon are aligned
and there is a flow of information commensurate with unaided human
perception so it is a communication technology, or more precisely a
material communication technology. With the telescopic gun sight the
flow of information is greater than the power of the human eye so
it is what Virilio (2000b, p. 83) calls 'an immaterial communication
technology' and for him the first such tool. Although a physically deadly
force does not flow from the camera to its target, it still shoots pictures
of events, freezes them in frames and by doing so kills them with a
metaphysically deadly force.

The shift to the rapid-firing gun is the early stages of a trend that
Virilio (1989b, p. 1) sees increasingly 'in industrialised warfare, where
the representation of events outstripped the presentation of facts, the
image was starting to gain sway over the object, time over space'. The
representation of events supersedes not only the presentation of facts
but also the event itself. The representation of an event exercises the
power of time over space, a magical power, but a magical power ascribed
to and embodied in the technologies of representation and the machines
of communication.

The modern magic wands of communication technologies with power
over time and space were used in war with deadly effect, and were even
developed for use in war in the first place. Indeed, for van Creveld
(1985, p. 104), communication technology is 'the stepchild of war' and
'the father of invention'. More pointedly, for Virilio, 'the war-machine
is not only explosives, it's also communications, vectorization' (Virilio
and Lotringer, 1983, p. 20; see also Virilio, 1995, p. 54). Communication
technologies are not so much the stepchild of war as the Siamese twins of
armament technologies: joined together in conception and birth, they
can only be separated by messy, surgical intervention with the survival
of one or both not guaranteed.

Benjamin (1973b, p. 84; 1999b, pp. 731–732; 2002, pp. 143–144) was
acutely aware of the devastating effects of war on human communication, if not of the close connection between communications technologies and war. He described how

> beginning with the First World War a process began to become
> apparent which has not altered since then and continues to this day.
> Was it not noticeable at the end of the war that men returned from
> the battlefield grown silent – not richer, but poorer in communicable

experience...For never has experience been contradicted more thoroughly than strategic experience by tactical warfare, economic experience by inflation, bodily experience by mechanical warfare, moral experience by those in power. A generation that had gone to school on a horse-drawn streetcar now stood under the open sky in a countryside in which nothing remained unchanged but the clouds, and beneath these clouds, in a field of force of destructive torrents and explosions, was the tiny, fragile human body.

Strategic experience involved looking over a limited battlefield, and/or maps of it, and gaining thereby an overall sense of the play of forces whereas tactical warfare is more piecemeal and involves looking over aerial photographs of many small areas of overall complexity and extent – Europe as battlefield.

Commenting on Benjamin's account of the First World War, Galvin (1994, p. 185) suggests that for him 'the most significant fact of the modern age' was 'a fundamental gap between a technological scale of events and human experience unmediated by such technologically advanced systems'. This gap meant, for Benjamin (1973b, p. 146; 1994, p. 564), that 'that reality of ours which realizes itself theoretically, for example, in modern physics, and practically in the technology of modern warfare...can virtually no longer be experienced by an *individual*'. Increasingly, however, hypermodernity is characterised by human experience mediated by even more technologically advanced systems producing an even wider gap between a technological scale of events and individual human experience; between the powerful forces of communication technologies and 'the tiny, fragile human body'; between the electronic speed of the hardware and the complex sophistication of the software on the one hand and the organic rhythms and intuitive imaginings of the wetware on the other; between the quantity of information on offer and the difficulty for humans to differentiate its quality.

Not only has technology, as Robins and Webster (1999, pp. 153 and 153–158 for a survey) put it, 'long been a key element of war', and not only have military and industrial power been interconnected, especially in the United States (see Smith, 1985, p. 4), but also communication technologies in particular have been used in war with deadly effect. Communication technologies have also been used by the media to record and report war, sometimes critically, sometimes not; sometimes under the control of the military, sometimes not (for a critical survey see Young and Jesser, 1997). The use of communication technologies destructively in fighting wars and critically in reporting wars encapsulates their paradoxical nature. They are also the means by which the military and the media are married as a mutually dependent couple in which one is not possible without the other, however defensive or manipulative the military, or critical, censored or sycophantic the media may be. Communication technologies are the matrix for the military and the media.

5
The Hell of Images: Cinema Paradiso

Film, for Benjamin (1986, p. 55) writing before the advent of television, was 'one of the most advanced machines for the imperialist domination of the masses'. Yet the cinematic machine is not only outside 'the masses' dominating them (us?) from without, but also inside our own heads dominating them/us from within. We are complicit with the machine and collude in our/its domination: 'the machine is us', as Haraway (1985, p. 99) says. The cinematic institution, for Metz (1982, p. 7), 'is not just the cinema industry . . . , it is also the mental machinery – another industry – which spectators "accustomed to the cinema" have internalised historically and which has adapted them to the consumption of films'. 'The mental machinery' of cinema, as Neale (1985, p. 1) puts it, is 'an apparatus for the production of meanings and pleasures, and as such involves aesthetic strategies and psychological processes'. There is no distinction between the machine and us insofar as we are constituted as mental beings. The machine is us mentally (see Giblett, 1996, Figure 1). Cinema is a sublime communication technology of mind–body dualism which, in Carey's (1989, p. 206) words, 'locates vital energy in the realm of the mind' and not, by implication, in the realm of the body.

Vertov vertigo

There is a distinction between the machine and us insofar as we are bodily beings. The machine replaces physical work and activity, but it is a mental entity before it is a physical one, an idea before an invention, and it is internalised in our minds rather than in our bodies. This internalisation has become naturalised over a hundred years of cinema-going, but it was entirely new for the first cinema-goers and cinematographers, or more precisely 'camera-eye men'. Dziga Vertov (1984, pp. 14–15), 'the man with a movie camera,' is usually seen as the apostle and evangelist (and no apologist) for the camera as eye and for 'the use of the camera as a kino-eye, more perfect than the human eye'. Not that the human eye is imperfect, just that the camera is more perfect. Vertov (1984, p. 17) was also the polemicist and proponent of the body as vehicle for the camera. Both camera as perfected eye and body as camera-vehicle come together in the boast that 'I am kino-eye, I am a mechanical eye. I, a machine, show you the world as only I can see it.'

Vertov (1984, p. 6) elaborated the implications of this mind–machine–eye–camera–body combo in his 'We: Variant of a Manifesto', in which he announced that 'we call ourselves *kinoks* ("cinema-eye men") as opposed to "cinematographers"'.

Kinoks, for Vertov (1984, p. 7), 'satisfy 'man's ... desire for kinship with the machine' as 'we bring people into closer kinship with machines' (Vertov, 1984, p. 8). 'Man' and machine are constituted as members of the same family, or at least the machine is a long-lost relative with which 'man' wants to establish his filial relations. In order to overcome the alienation of 'man' and machine, the kinok wants to acknowledge that the machine is related to and produced by 'man'. And he wants to enact that kinship in the act of filming.

Yet in the process the machine is anthropomorphised, or at least animated, and rhapsodised. To become machine, 'man' has to animate machine. Vertov (1984, p. 9) gives an 'hurrah for the poetry of machines, propelled and driving the poetry of levers, wheels, and wings of steel, the iron cry of movements, the blinding grimaces of red-hot streams'. The machine is figured as an animated body (in two senses, animal body and moving body, not static body of ore) that emits terrifying screams and burns out the eyes. The machine is an instrument of the law of the Father that operates through terror and the threat of castrating the phallic power of sight.

The machine, for Vertov (1984, p. 8), even has a soul and the poetry of machines is in the business of 'revealing the machine's soul'. Fulfilling 'man's desire for kinship with the machine and its soul' means transforming 'man' into machine, albeit an animated, rhapsodised, even sublimated, machine. For Vertov (1984, p. 8), '*The new man*, free of unwieldiness and clumsiness, will have the light, precise movements of machines.' Kinoks sublimate the heavy and solid, cumbersome and clumsy human body into the light and ethereal, dexterous and precise movements of the machine. For Vertov (1984, p. 8) and his fellow kinoks, '*our path leads through the poetry of machines, from the bungling citizen to the perfect electric man*'.

Unlike Whitman in his *Leaves of Grass*, who merely *sang* the body electric (see Benthall, 1976, pp. 13 and 163–171), Vertov in his *Man with a Movie Camera* and as the man with a movie camera *made* the body electric just as the cinema in general electrified the human body (see Christie, 1994, pp. 64–87). Photography enacted the Promethean desire to snatch fire from the gods in the heavens and bring it down to humans on earth; cinema achieved that desire in what Christie (1994, p. 65) calls, using Mary Shelley's subtitle of *Frankenstein*, 'the Modern Prometheus'. Electricity is the modern fire; cinema is one of its vehicles for coming down to earth from the heavens (radio is another as we shall see in a later chapter); the cinematographer (and the star as we shall see later in this chapter) is the body electric.

Similarly, electricity, for Villiers de l'Isle Adam (1981, p. 149; see also p. 77), was 'the spark bequeathed by Prometheus'. In his novel of the 1880s, *Eve of the Future Eden*, a fictionalised Edison 'with the sublime aid of light' creates what is variously described as 'a magneto-electrical entity', 'a new electro-human creature', 'an electro-magnetic creature' or simply 'Andreid' (de l'Isle-Adam 1981, pp. 67, 72, 113 and 180), the first fictional Android, called Hadaly (meaning the Ideal). In the whirring armatures and phonographic speech of 'the magnetic-metallic organism of Hadaly' (de l'Isle-Adam, 1981, p. 92), Edison claims he will 'bring illusion down to earth ... and force the Ideal to show itself for the first time to your senses, palpable, audible, and materialised' (de l'Isle-Adam, 1981, pp. 72–73).

What Shelley and l'Isle-Adam only dreamed and wrote about cinema was to realise. Cinema brought dead matter to life on the screen. Hadaly as 'a regal vision machine, almost a creature, a dazzling simulacrum' (de l'Isle-Adam, 1981, p. 97) was the forerunner of the cinematic cathedral and televisual cathode-ray 'vision machine' (Virilio, 1994b) of cinema and television. She is a dead living star. Hadaly, the Andreid, 'knows no life, no disease, no death. She is above all imperfections and all servitudes' orbiting in the sublime company of heavenly bodies where 'she keeps her ethereal beauty' pure and unsullied, like the dead living star on the silver screen (de l'Isle-Adam, 1981, p. 177).

The modern communications technologies of photography and cinema employed the light of the sun, the fire in the sky. Cinema later employed the artificial light of the modern fire of electricity to project film onto a screen. Modern industrial technologies unbound Prometheus (see Landes, 1969, especially pp. 24 and 284) to release the power of heat in thermodynamic technologies (steam and internal combustion engines), and of light and electricity in communication technologies (photography, cinema, telegraphy, radio and television). All of them entailed what Landes (1969, p. 24) calls 'mastery over nature' and 'mastery of the environment' rather than mutuality with them (see Giblett, 2004).

Electricity is the connection between the body of 'man' with its nerves and the cinema with its light machines and displays. And its nerves. For Vertov (1984, p. 8), 'cinema's unstrung nerves need a rigorous system of precise movement'. It needs nerves of steel, nerves of electrical impulses. Cinema needs a grid of wires to circulate power and energy and to control electrical circulation through command, control and feedback. Cinema needs and uses what Vertov (1984, p. 7) called 'electricity's unerring ways'. Electricity constitutes cinema as machine and the human body as electrical machine. Electricity becomes the model for cybernetic control of the human body by way of the nervous system. With cinema, human beings become a cybernetic organism, a cyborg. The first cyborg was the cinematographer and cinema-goer. Previously in the physiology of Leonardo da Vinci and René Descartes the human body was a machine; in Vertov's physiology the human body becomes an electric motor.

Previously for Leonardo da Vinci the earth also was a machine; with Vertov the whole earth is an electric motor. For Vertov (1984, p. 8),

openly recognizing the rhythm of machines, the delight [*sic!*] of mechanical labour, the perception of the beauty of chemical processes, WE [*sic*] sing of earthquakes, we compose film epics of electric power plants and flame, we delight in the movements of comets and meteors and the gestures of searchlights that dazzle the stars.

Searchlights, as Bailes (1980, p. 67) puts it, 'dramatically symbolized the new technology [of] electrical illumination' and were first used in war during the Boer War (as he goes on to relate). They were also used in the lead up to the Second World War when Albert Speer created Zepplinfield composed of 'walls of light' using 150 searchlights (see Virilio, 1989b, p. 78). Zepplinfield and cinema both produce 'photo-murals'. The play of light is fleeting, evanescent but fixed in virtual walls that immobilise the spectator. Cinema is a sublime communication technology as it deals with creative, large-scale events and transforms them into soaring and transcending walls of light, into virtual walls.

'Man' becomes cybernetic organism in what Vertov (1984, p. 120) calls 'cinema's belly'. Cinema is animated as the Belly of the Beast (Leviathan, the monumental State) and as the belly (mechanical womb) of a Bachelor Machine for Bachelor Births (see Giblett, 1996, p. 50, n. 7). Carrouges (1975, p. 21; see also Carrouges, 1954; Certeau, 1986) defines a bachelor machine as 'a fantastic image that transforms love into a technique of death'. Out of this machine, new life is created from dead matter just like Shelley's Frankenstein, who had 'a fervent longing to penetrate the secrets of nature' and 'show how she works in her hiding places' (Shelley, 1992, pp. 39 and 47; see also p. 53). Thinly veiled sexual and gynecological metaphors are used to figure his burning desire to know in a combination of both the biblical (sexual) and the epistemological senses of the word. Out of this combined act of rape and insemination of nature, Frankenstein and the kinok would be impregnated and then with 'intense labour' (Shelley, 1992, p. 39) would 'bestow animation upon lifeless matter' (Shelley, 1992, p. 51) and bring to light the dark workings of nature, and life out of dead matter (though not out of nothing unlike God).

A bachelor machine is basically a masturbatory machine. Cinema as a masturbatory machine is graphically depicted in Fellini's film *Amarcord*, in which adolescent boys jerk off in a bouncing parked car intoning the names of Italian film stars in a sexual litany. Similarly, films, for Dylan Thomas, are 'our eunuch dreams, all seedless in the light' (quoted by Chanan, 1980, p. vi). Cinema represents the impossibility of reproduction by the cinema-goer. 'He' is castrated (by whom? the father-state? the paternal corporation? both?) and rendered infertile, incapable of procreation. So he goes to the cinema to simulate birth out of his head. The womb is sublimated into the head that gives birth to the brainchild. Yet unlike Zeus who gave birth to Pallas-Athena out of his head (see Michael Meier's engraving in Theweleit, 1989, p. 345), and the photographer who gives birth to a new world out of his box brownie, the kinok gives birth to a new woman out of the eye/camera lens.

The lens of the camera-eye, for Vertov, is a sphincter that opens and shuts, not only to receive light and record still-life (still-birth), but also to give birth and make new life (for an illustration of the lens see Vertov, 1984, p. 15). Yet the camera is an orally and anally sadistic monster that takes life and gives death at the same time as it gives birth. The lens of the camera-eye is an oral and anal (or more precisely, cloacal) sphincter. It excretes and gives birth through the same hole, a cloaca. The camera is an instrument of visual sadism. If photography, as Benjamin argued (1973b, p. 221; 2002, p. 102; 2003, p. 253), represented the victory of the eye over the hand in pictorial reproduction (as we saw in the previous chapter), cinema, for Vertov, represented the victory of the eye over the brain, not only in pictorial reproduction but also in non-biological human reproduction, in mechanical pictorial production.

The kinok is not only a Greek god on Mount Olympus who gives birth like Zeus to a new woman in the star, but also the Judeo-Christian God in the Garden of Eden who creates a new man, like Shelley's *Frankenstein*. Vertov (1984, p. 17) claims that 'I am kino-eye, I create a man more perfect than Adam.' Whereas the God of biblical theology created Adam by breathing life into dust, and promised to create a new Adam through salvation and resurrection, cinema creates a new Adam of the spectator through the power of the eye, and resurrects dead images in the new Eve of the star. The power of the eye has superseded the power of breath, just as the power of breath had superseded the power of water (see Giblett, 1996). The power of the eye in filiarchy, the rule of the sons in modernity, especially in industrial capitalism, superseded the power of the breath (spirit) in patriarchy, the rule of the fathers in pre-modern and early modern, pastoral and agricultural society. This power, in turn, had superseded the power of water in the matrifocal rule of gylanic societies (see Giblett, 1996; 2004, chapter 1).

The cinema-eye had the power to give birth to a new kind of man as it had the power to produce images combined in montage. Montage could combine an image of a human eye with an image of a camera lens to show the two as one (see Vertov, 1984, p. 15). Vertov (1984, p. 17) claimed that 'through montage I create a new, perfect man'. Cinematic montage combined two or more shots in one. Montage puts events together to compress space and time. The mechanical eye, the camera, according to Vertov (1984, p. 19), 'experiments, distending time, dissecting movement, or, in contrary fashion, absorbing time within itself, swallowing years, thus schematizing processes of long duration inaccessible to the normal eye'. The movie camera concertinas time, both compressing it within a moment and expanding it across duration.

Montage enabled the human eye to see in ways that were hitherto impossible. For Vertov (1984, p. 19), 'the mechanical eye, the camera, rejecting the human eye as crib sheet, gropes its way through the chaos of visual events, letting itself be drawn or repelled by movement, probing, as it goes, the path of its own movement'. Montage does not create order out of chaos, but makes a pathway through it. The camera is a kind of masculinised mechanical speculum that undertakes an active internal examination of feminised, passive visual events.

The cinema camera is a mobile eye soaring vertiginously into the sublime heights and plunging vertiginously down from them like a plane. Vertov (1984, p. 17) claimed in probably his most famous and oft-quoted statement that

> Now and forever, I free myself from human immobility, I am in constant motion, I draw near, then away from objects, I crawl under, I climb into them. I move apace with the muzzle of a galloping horse, I plunge full speed into a crowd, I outstrip running soldiers, I fall on my back, I ascend with an airplane, I plunge and soar together with plunging and soaring bodies. Now I, a camera, fling myself along their resultant [*sic*], manoeuvring in the chaos of movement, recording movement, starting with movements composed of the most complex combinations.

The kinok is mobile, but the spectator is immobile, or at least only virtually mobile. Cinema enacts what Kirby (1997, pp. 2 and 3) calls the paradox of 'simultaneous motion and stillness...In cinema...the perceptual illusion of movement is tied to the physical immobility of the spectator.' The spectator moves mentally but not corporeally. The cinema is a vehicle for a seated passenger like the panorama, diorama and railway before it and the car, television and computer after it. The cinema-goer travels across space and up and down the class scale. For Rosalind Williams (1982, p. 79), 'whether the distance is geographic or social, film allows the pleasures of mobility' – with few of the discomforts. The same visual pleasures are enjoyed irrespective of whether the passenger travels away from home in the railway carriage and car or stays at home watching television, or close to home in the cinema.

'What was the good of moving', complained Huysmans (1959, p. 142) des Esseintes in the motto of hypermodernity, 'when a fellow could travel so magnificently sitting in a chair'. Even when the chair moved with the car or the train, a fellow did not move his limbs and body but was moved about. A fellow could travel magnificently sitting in a chair irrespective of whether the chair moved or not. He may even be 'carried by a vehicle into a sphere where sublimated sensations would arouse within him an unexpected commotion' (Huysmans, 1959, p. 180). This was des Esseintes' Decadent ideal for a work of art, but it could also be the apogee for the aestheticisation of everyday life by the real and virtual vehicles of modern transportation and communication technologies. Everyone and everyday life become decadent and not merely aestheticised in and by the vehicles of virtual mobility like cinema and television.

Cinema is mental machinery. The machine is mental to the point that Benjamin (1973a, p. 132; 1973b, p. 177; 2003, p. 328) argues that

> Technology has subjected the human sensorium to a complex kind of training...In a film, perception in the form of shocks was established as a formal principle. That which determines the rhythm of production on a conveyor belt is the basis of the rhythm of reception in the film.

The early 'cinema of attraction' (Gunning, 1986; 1989) was a cinema of shock ('rapid cutting, multiple camera angles, instantaneous shifts in time and space') (Wolin, 1982, p. 233). The shock of the new came at the price of the loss of the old, the traditional, the auratic. For Benjamin (1973a, p. 154; 1973b, p. 196; 2003, p. 343), 'the price for which the sensation of the modern age may be had [is] the disintegration of aura in the experience of shock'. The instantaneity of the shock administered by the moving pictures of cinema (and later television) disintegrated the protraction of the aura. Just as shorter exposure times and brief posing for the snapshot brought about the demise of the lengthy sitting for the auratic early photos, so the production of shocks on the assembly line of the cinema diminished the auratic presence of objects. Just as the trace and aura for Benjamin are distinct in photography, so are shock and aura in cinema.

The cinema of shocks arose out of advertising and the new pictorial representationalism instigated by print and lithography. For Benjamin (1979, p. 89; 1996, p. 476),

> today the most real, mercantile gaze into the heart of things is the advertisement. It abolishes the space where, and tears down the stage upon which, contemplation moved and all but hits us between the eyes with things as a car, growing to gigantic proportions, careens at us out of a film screen. And just as the film does not present furniture and facades in completed forms for critical inspection, their insistent, jerky nearness alone being sensational, the genuine advertisement hurtles things at us with the tempo of a good film.

The early 'cinema of attractions', such as *Arrival of a Train at a Station* that depicted just that, produced not only shock but also such terror that audiences reportedly ran screaming from the cinema. Such a response was indicative not merely of a naïve realism in the audience that mistook the pictorial representation of the thing for the thing itself, that collapsed realism and reality, but also of a terror at a thing grown to such monumental proportions that it towered above and threatened to engulf the minuscule members of the audience, the terrifying disproportion between the sublime machine of the train in the cinema and what Benjamin (1973b, p. 84; 2002, p. 144; 1999b, pp. 731–732) calls 'the tiny, fragile human body'.

The power of 'the movies' lay not just in moving pictures, in the ability of the cine-camera to take pictures continuously (unlike the photographic camera), but also in the moving camera, in the ability of the camera to move and take pictures whilst it is doing so. Benjamin (1999b, p. 17; see also 1973b, p. 238; 2002, p. 117) commented that

> with film, a *new realm of consciousness* comes into being. To put it in a nutshell, film is the prism in which the spaces of the immediate environment – the spaces in which people live, pursue their avocations and enjoy their leisure – are laid open before their eyes in a comprehensible, meaningful and passionate way. In themselves these offices, furnished rooms, saloons, big-city streets, stations and factories are ugly, incomprehensible and hopelessly sad. Or rather, they were and seemed to be until the advent of film. The cinema then exploded this entire prison-world with the dynamite of its fractions of a second, so that now we can take extended journeys of adventure between their widely scattered ruins.

As the adventure-journey through ruinous landscapes is the hallmark of Romanticism, cinema created a romanticised, aestheticised and touristic landscape out of the ordinary and everyday. The cinema-goer is a tourist of the everyday; cinema is tourism of the everyday, and tourism for the everyday. The cinematograph, Virilio (2000a, p. 23) argues,

> was a substitute for human vision which not only flouted time (thanks to the illusion known as persistence of vision), but also flouted the distances and dimensions of real space. The cinema was, in fact, a new energy, capable of carrying your gaze to other places, even if you yourself did not move.

The gaze was carried to other places whilst the body was still sitting in one place. The gaze was moblised and the body immobilised, even demobilised.

Whereas in landscape painting the Claude glass compressed the country landscape within a concave mirror to make it into a static, painterly object, cinema exploded the city landscape with retinal retention into a continuous stream of fragmentary images. Whereas the Claude glass implied and addressed a single viewer viewing from a single point of view, cinema addressed and positioned an audience seeing the same thing from the same point of view. For Benjamin (quoted by Virilio, 1994b, p. 21), 'cinema provides *matter* for simultaneous collective reception'. This statement implies the question of which matter? What sort of matter? Virilio (1994b, p. 21) answers this question by stating that 'the *matter* provided and received in collective, simultaneous fashion by cinemagoers is light, the speed of light', nor the matter of the solid object filmed for that has been sublim(at)ed into light, become a virtual object and then deposited in the sublimate, the solid object, of film.

For Virilio (1991a, p. 18), Marey, the inventor of the chronophotographic gun, gives light the leading role, 'makes it the leading lady in the chronophotographic universe'. Marey makes light, in other words, into the star, and cinema makes the star into light, or at least, as Virilio (1991a, p. 54) puts it, 'as diaphanous as if the light was pouring through her flesh'. In the cinema, for Virilio (1989b, p. 38), 'everything visible appears to us in the light'. That which is (in) the dark is not visible, at least to the cinematic eye of reason. What is visible can be appropriated; the invisible cannot be appropriated. With the development of what Comolli calls 'machines of the visible' (1980, pp. 122–123) 'the whole world becomes visible at the same time that it becomes appropriatable'. The invisible is by definition inappropriable, and inappropriate.

In order to be appropriatable, the world first had to be not only seen but also known. Knowing was reduced to seeing; seeing was knowing; to see was to know in a technological reduction of God's function. With cinema, Virilio (1989b, p. 4) argues, 'in a technicians' version of an all-seeing Divinity, ever ruling out accident and surprise, the drive is on for a general system of illumination that will allow everything to be seen and known, at every moment and in every place'. Why? Because one is in every moment and place; one is omniscient because omnipresent. Abel Gance (quoted by Virilio, 1989b, p. 26) defined cinema as 'magical, spell-binding, capable of giving to the audience, in every fraction of a second, that strange sensation of four-dimensional omnipresence cancelling time and space'.

Or at least as Christie (1994, p. 12) more precisely puts it, 'moving pictures gave early audiences an expanded and concentrated experience of time'. The railway supposedly annihilated space; the telegraph supposedly annihilated time; photography supposedly compressed time and space together; but cinema, for Gance, cancelled them altogether. As Christie (1994, pp. 10 and 17) goes on to suggest, 'the magic was now in the movement... From the carriage window to the screen was an easy transition [as it was later to the car windscreen and computer monitor]. It's tempting to say that sixty years of railways had prepared people to be film spectators', as well as car drivers, television viewers, online gamers, e-consumers and Internet surfers.

In looking backwards from the railway to cinema, Kirby (1997, p. 2) argues that 'the cinema finds an apt metaphor in the train, in its framed, moving image... the "annihilation of space and time"'. It was 'an instrument for conquering space and time'. For her 'both the railroad and the cinema wiped out pre-existing ideas and experiences of time and space' (Kirby, 1997, p. 48). In looking backwards and tracing the genealogy of communication technologies, cinema is the direct descendant of the railway as much as it is of photography. Cinema = railway + photography. In looking forward from the railway to cinema, Cubitt (1998, p. 29) argues that 'the railway train is closely linked to the emergence of cinema, its windows offering the kind of moving views for which the cinematograph was to become famous in the 1890s'. Both the railway and the cinema converge in the windows of the computer hardware and software (Windows®) for which computers and the Internet became famous in the 1990s.

Star

Yet despite this emphasis on light and seeing, cinema has been critiqued as a place of darkness and blindness. The early cinemas, Virilio (1989b, p. 31) relates, were regarded as 'deconsecrated sanctuaries in which, as Paul Morand put it, the public sensed the end of the world in an ambience of profanation and black masses'. Cinema is a secular sacrament that commemorates the 'death' of the actor in filmed images and their transformation into screened images. These images, in turn, are transubstantiated into the heavenly body of the star, the saints of the cinema cathedrals. Virilio (1989b, p. 38) points out the similarity and succession between 'the stained-glass windows of old cathedrals' and what he calls 'the American cinema cathedrals' (p. 31), or 'the cinema-cathedrals of the military state' (p. 38). In this succession and this fascination with coloured light the Paris Expo of 1900 should be included with its Palace of Electricity, whose façade was described by a contemporary reviewer as 'a stained-glassed window of light' (cited in Rosalind Williams, 1982, p. 85; see also Schivelbusch, 1988, p. 73).

In the cinema cathedrals a formerly state, and latterly a corporate, cult of death is enacted in which the body of the star is projected and sacrificed on the screen. Cinema cathedrals are also tombs in which the image of the star is immured and the womb from which the image is resurrected. The cinema, for Dylan Thomas, is a place of 'the dark brides, the widows of the night...the shades of girls' (quoted by Chanan, 1980, p. vi). A photograph, for Neale (1985, p. 8), 'embalms the ghosts of the past; film brings them back to life' after it has put them to death, for 'film is exactly a putting to death, the demonstration of "death at work" (Cocteau's "la mort au travail")' (quoted by Neale, 1985, p. 55). A photograph preserves or even gives life to the dead so that they become the living dead, and it kills the living so that they are rendered and become not merely dead but the dead living.

Film resurrects the living dead and the dead living by making them move. Cinema for early commentators was 'pictures with people coming alive' and moving 'exactly as though they were alive' (quoted by Chanan, 1980, pp. 3 and 4). Stars were, in the words of Antonin Artaud (quoted by Virilio, 2000a, p. 30), 'creatures *condemned to die alive*', and the star is '*a ghost that you can interview*, said Michel Simon' (quoted by Virilio, 1991a, p. 54). Hollywood, for Virilio (1986a, p. 29; 1991b, p. 26), is 'the city of living cinema where...the living and the living dead mix and merge deliriously' to the point of producing the dead living star.

The state is war by other means; 'the cinema is war pursued by other means', as Virilio might have said (Redhead, 2004, p. 108), or more precisely, the cinema is war by another medium; the star is a war bride married to war. The star system arose just before the First World War. According to Christie (1994, p. 87), Asta Nielsen, who began making films in 1910 and starred as 'a beautiful but modernly ironic *femme fatale*' in two films in 1913, was 'the first truly international female star'. The star system, for Virilio (1989b, p. 30), 'only really triumphed after 1914 in the new cinema industry'. Why? Because, for Virilio (1989b, p. 23), 'the star system and the sex symbol were the result of that unforeseen perceptual logistics which developed intensively in every field during the First World War'. The landscape of the battlefield and the body of the star were subjected to the same cinematic techniques and produced a similar optical effect. With the star, cinema produced the film actor as body, just as cinema produced the earth as body to be filmed. For Virilio (1989b p. 26), 'once the star had been named "body" and her picture painted on bombs and bombers, this body with no stable dimensions would soon be offered up in "fragments" to the audience, in a repetition of the perception of the military voyeur' poring over his aerial photographs of the battlefield like a connoisseur of eroticism or a consumer of pornography.

The fragments of the star's body in the frames of the film are both fetishes of her whole body and wafers and wine in the sacrament of the black mass of the cinema cathedrals in which the body of the star is transubstantiated into dead living images. The star is a mediating category between solid body and insubstantial images. The star system after 1914, Arnheim argues, 'oscillated in zones higher than the universe of practical things and lower than the disembodied forces that animate such things' (quoted by Virilio, 1989b, p. 41). Oscillated, in other worlds, in the ethereal realm of the sublime between the solid matter of human bodies and the heady heights of abstraction, ideality and theory. The oscillating zone of the star (and the young virginal woman, the vestal virgin) is the sublime counterpart to the slimy quaking zone of the swampy Great Mother (see Giblett, 1996).

What Virilio (1989b, pp. 42, 33–34) calls 'the blazing of the female star' (the sun in each image) beginning around the time of the First World War occurred at a time when

> a whole series of exchanges and transfers of power take place between male warriors and logistical spouses [war-brides] – that is to say, between natural reproduction of the old gynaeocracy (in which kinship is mostly matrilinear) and all the techniques for preserving and reproducing the new City-State . . . *a travesty of birth, a field of death dressed up as life* (his emphasis).

The star is a war bride transferring power between the Great Mother and Father Law, between swamps and the city, between the Magna Mater of living beings and the Matrix of dead things, between the womb of life and the tomb of death (see Giblett, 1996, chapter 2; 2004, chapter 1). Virilio (1989b, p. 42) claims that 'at that time [of the First World War], women played an ambiguous role in narratives, emanating from a rather uncertain and dangerous world of water and impenetrable, magical forest'. This watery world of northern lights and sacred swamps had to be penetrated and drained by phallic heroes (see Giblett, 1996) using the technologies of empire such as theodolites and cameras, guns and railways. A spatial metaphysics and poetics had been constructed in which, as Kern (1983, p. 242) puts it, 'low suggest[s] immorality, vulgarity, poverty, and deceit. High is the direction of growth and hope, the source of light, the heavenly abode of angels and gods', and stars.

The star is quintessentially a female body. Of course, stars can be and are male, but they have been feminised as their bodies have been fetishised, cut up into pieces, morcellated, fragmented and endowed with life, the dead living. Cinema was to animate and locomote the female body to the point where, as Michelsen (1984, p. 20) puts it, it was constituted '*as* cinema', 'as the very site of cinema's invention', and cinema's continual re-invention and re-animation of the female body on the screen.

The heavenly bodies of cinema stars were described by Pirandello (quoted by Virilio, 1989b, p. 15; see also Benjamin, 1973b, p. 231; 2002, p. 112; 2003, p. 260) as being 'so to speak subtracted, suppressed, deprived of their reality, of breath, of voice, of the sound they make in moving about, to become only a dumb image, which quivers for a moment on the screen and disappears, in silence'. The body of the star is reduced to mute visual image that burns with a fleeting, incandescent light, and then dies. Their unique presence in time and space, their address to the ear, the aural, their aura, has gone, especially and quite literally in the early silent films.

Benjamin (1973b, p. 231; 2002, p. 112; 2003, p. 260) goes on to elaborate on the situation outlined by Pirandello: 'for the first time – and this is the effect of the film – man [*sic*] has to operate with his whole living person, yet forgoing its aura. For aura is tied to his presence; there can be no replica of it.' The star system is, for Benjamin, the precise counterpart and exact compensation for this loss of aura in film. Benjamin (1973b, p. 233; cf. 2002, p. 113; 2003, p. 261; see also Caygill, 1998, pp. 108–109) argues that

> the film responds to the shrivelling of the aura with an artificial build-up of the 'personality' outside the studio. The cult of the movie star, fostered by the money of the film industry, preserves not the unique aura of the person, but the 'spell of the personality', the phoney spell of a commodity.

Aura is the sacral quality of an object present in time and space; the star is the fetishised quality of a commodity cancelling time and space. The star is sublimated into a transcendental realm between solid matter and heady ideality; aura is desublimated into an immanent realm between solid matter and slimy traces.

The star made art into commodity and paved the way for art to be made into politics. As Johnson (1988, p. 6) eloquently puts it, Benjamin 'pointed to the creation of movie stars and the building up of the cult of the personality by film studios as an attempt to replace the desiccated aura of the art object with the artificial aura of the commodity'. Film studios sowed the wind of the movie star only to reap the whirlwind of the cult of personality in the totalitarian dictator, Fascist and 'Communist' (State Capitalist) alike. For Benjamin (1973b, pp. 243–244; 2002, pp. 121–122; 2003, pp. 269–270), 'the logical result of Fascism is the introduction of aesthetics into political life' to which 'Communism responds by politicising art.' The artificial aura of the commodity superseded the dried-up aura of art, and religion, and substituted commodity fetishism for sacrality, dead matter for living being. Capitalism aestheticises commodities. The modern conquest of the world as picture and the hypermodern conquest of the world as commodity ultimately come together.

Yet the visual fragmentation, fetishisation and commodification of the female body had already taken place at least a decade before the rise of cinema and the birth of the star. Between 1880 and 1910 what Leach (1993, pp. 51 and 50) calls the 'new advertising pictorialism' using 'new kinds of color and light' arose. From the 1880s, Raymond Williams (1968, p. 26; see also 1980a, pp. 176 and 179) argues, 'new kinds of display advertising began to break into the Press, at a time when changes in marketing and the development of the retail trade were changing the whole basis of advertising'. There a shift was not only from classified advertising to display advertising in what has come to be called the 'Northcliffe Revolution' based on what Williams (1958/2001, p. 18) calls 'the new mass advertising of the 1890s', but also from alphabetic text to photographic representation. Advertising was becoming what Williams (1980a, pp. 184 and 187) also calls 'the official art of modern capitalist society' and 'the magic system' in which meanings and values were ascribed to commodities. These developments linked up with the development of electric lighting and its use in department stores and shop windows (see Schivelbusch, 1988, pp. 143–154; Williams, 1982, pp. 84–90). Together they produced new forms of display of commodities in the virtual spaces of advertising photographs and shop windows.

Both of these developments are well illustrated in Rudyard Kipling's (1904, pp. 220–221) story 'Wireless' set at night in a chemist shop lit up with electric light. The first-person narrator describes 'the seductive shape on a gold-framed toilet-water advertisement whose charms were unholily heightened by the glare from the red bottle in the window'.

For one of the characters whose eyes were 'bent' towards the ad 'the flamboyant thing was to him a shrine'. The female figure of the ad is reduced to a colourful object of desire and imbued with unholiness, a thing set apart and immured within an unholy shrine, a monumental memorial to the dead living maiden of the ad. Instead of the auratic object enclosed in an ornamental case, the flamboyant thing is buried in an unholy shrine. Aura has become flamboyancy.

The logistics of perception at work here is produced by electric lighting and photography that cinema was only to combine in the 'stained-glass windows', or screens, of its cathedrals. Kipling's narrator comments that 'there was, after all, a certain stained-glass effect of light on the high bosom of the highly-polished picture' (Kipling, 1904, p. 228). The electric light refracted through the stained-glass bottles of the chemist shop onto the photo of the scantily clad figure of the young woman in the advertisement produces a static tableau of commodification and fetishisation of the female body. Cinema was to animate and locomote the female body to the point where, as Michelsen (1984, p. 20) puts it, it was constituted '*as* cinema', 'as the very site of cinema's invention', and as cinema's continual re-invention and re-animation of the female body.

War

The cinematic logistics of perception produced not only the star as war bride nor only the aerial photography of the battlefield. It also produced the landscape as battlefield in the First World War, the first, and last, war conducted largely in trenches. Commenting on the landscape of trench warfare in the First World War, Virilio (1989b, p. 14) argues that 'to the naked eye, the vast new battlefield seemed to be composed of nothing – no more trees or vegetation, no more water or even earth, no hand-to-hand encounters, no visible trace of the unity of homicide and suicide'. This nothing landscape that was by no means utopic, a no place, or even atopic as place had been obliterated, was cinematic. For Virilio (1989b, p. 70), 'the landscape of war became cinematic' with the First World War. Why? Because it, like the star, was idealised and feminised.

Drawing on the terminology of Jünger's 1920 (2004, especially pp. 38–39; see also Barbusse, 2004) classic novel of the First World War, *Storm of steel,* the cinematic landscape of trench warfare was, for Benjamin (1999, pp. 318–319), the product of German idealism:

> it should be said as bitterly as possible: in the face of 'this landscape of total mobilisation,' the German feeling for nature has had

an undreamed-of upsurge. The pioneers of peace, who settle nature in so sensuous a manner, were evacuated from these landscapes, and as far as anyone could see over the edge of the trench, the surroundings had become the terrain of German idealism; every shell crater had become a problem; every wire entanglement an antinomy; every barb a definition; every explosion a thesis; by day, the sky was the cosmic interior of the steel helmet, and at night, the moral law above. Etching the landscape with flaming banners and trenches, technology wanted to recreate the heroic features of German idealism. It went astray. What it considered heroic were the features of Hippocrates, the features of death. Deeply imbued with its own depravity, technology gave shape to the apocalyptic face of nature and reduced nature to silence – even though this technology had the power to give nature its voice.

The camera mobilised the eye in and against the landscape and in the process produced a landscape of total war, the war of 'man' against nature, the war of 'the militarisation of science', as Virilio puts it (2003b, p. 85; see also Cornwell, 2003, pp. 16 and 377). The First World War of total war and the militarisation of science led to the Second World War of the almost total mobilisation of scientists and engineers. The Second World War, for Cowan (1997, p. 257), was 'the first war in which the nation's [United States'] scientists and engineers were almost completely mobilized for the war effort'. Despite their differences, warfare of the early and late twentieth century produced the same sort of landscape. For Gray (1989, p. 55), 'chemical, biological and nuclear (CBN) weapons make the killing fields of late twentieth-century battle as lifeless as the surface of the moon'. The landscape of warfare is 'like a lunar landscape, as Remarque (1929/1994, p. 40; see Giblett, 1996, Figure 1) said in *All quiet on the western front*.

Despite its ugliness, or perhaps because of it, this landscape was a feminised, albeit misogynist, landscape. Virilio (1989b, p. 22) argues that

women become *objective tragedy* in the wars from which they were excluded. The leer that the conquering soldier casts on a woman's now distant body is the same as that which he directs at a land turned into desert by war. It is also a direct antecedent of the cinematic voyeurism with which a director films the star as one does a landscape, with its lakes, contours and valleys.

Director Joseph von Sternberg, according to Virilio (1991a, p. 54), 'described the actress' face as a landscape, with its terrain, its lakes and valleys, over which the camera travels'. The close-up, Virilio (1991a, p. 57) argues later, makes the face of stars into 'battle-landscapes'. Yet not only the star's face but also her body became a (battle) landscape with its curves (hips), prominences (breasts) and valleys (cleavage, legs). The star became what Virilio (1991a, p. 58) calls 'the luminous spectre of the landscape'. Women were both excluded from war and became objective tragedy, became landscape, and included and became subjective sacrament, became star.

Aestheticised landscape turns into the landscape of trench warfare – both were feminised: the former as picturesque and pleasing prospects, the latter as the desert of the sublime in the horizontal. The sublime in the vertical (mountains, mountainous seas) was masculinised, whereas the sublime in the horizontal was feminised, whether it is in the deserts of the Australian war against nature (see Giblett, 2004, chapter 11) or in the wasteland of European trench warfare with its 'man-made' swamps. The metaphysics and poetics of high and low were translated into the heights and depths of the battlefield itself. The vertical dimension of the battlefield was altered as, in Kern's (1983, p. 310) words, 'the land was pushed down into the dugouts and mines and galleries of the trenches', the depths of the land below, the nether-lands.

6
Magician's Bower and Monstrous Mechanical: The Car as Communication Technology

If the telegraph separated communication from transportation, as Carey (1989) and others argue, the car rejoined them to produce a mobile communication and transportation device. The car, for Sachs (1992, p. 116) and Virilio (1991b, p. 21), is 'a means of communication', not only in the clichéd sense that it is a mode of transportation, but also in the sense that the driver (and his or her passengers) communicates with the landscape – built or otherwise, cultural and natural. He/she receives sensory impressions from it (albeit largely visual) and impacts physically on it. The car, for Friedberg (1993, p. xi), like cinema, is one of a number of 'machines that mobilize the gaze' and cinema spectatorship is 'a radical metaphor for the windshield' of the car. Similarly, for Virilio (1988, p. 188), 'what goes on in the windshield is cinema in the strict sense' of a moving image projected onto a screen, the windscreen in the case of the car. Finally, for Christie (1994, p. 22), 'both motoring and cinema offered the satisfaction of seeing the world "whizz by"' in a common visual sensation of speed.

The driver also communicates with other drivers, albeit in a highly circumscribed manner through the limited language of gesture, horn or lights or via his or her car's occupation of space. Yet at the same time that the car and the cinema are mobilising the gaze, they are both immobilising the rest of the body, or abstracting the gaze from the bodily bearer of looking. For Cubitt (2000, p. 131), it is ironic that 'the automobile and the cinema alike produce an immobile spectator of action, not a mobile participant'. Both are vehicles for seated passengers.

If the telegraph separated time from space, as Carey argues, also by enabling communication to proceed faster than transportation, the car rejoined them. The driver travels through space at the same speed, in the same time, as the car. The car has power over space and time yet its power is paradoxical. The car liberates and captivates at the same time: the car liberates the driver from the private sphere of the home, office or factory into the public sphere of the street, the road or the off-road, at the same time as it captivates the driver in the car's private sphere of mobile indoor space. The car, in J. G. Ballard's novel (1995, p. 199) *Crash*, is 'a magician's bower', whereas for Sachs (1992, p. 146) 'cars are "communication prisons"'. Rather than cars being one or the other (either/or), cars are both liberators and captivators (both/and). The car encapsulates the paradox of modernity: mastery of space and time, and slavery to both of them too.

Mobile privatisation, private mobilisation

The car is an instance of what Raymond Williams (1985, pp. 188 and 189) called 'mobile privatisation' in which the driver experiences movement within what he described as 'the conditioned atmosphere and internal music of this windowed shell'. Two decades ago he maintained that

at most active social levels people are increasingly living as private small-family units, or, disrupting even that, as private and deliberately self-enclosed individuals, while at the same time there is a quite unprecedented mobility of such restricted privacies. In my novel *Second Generation* (1964) [Williams goes on to relate] I developed the image of modern car traffic to describe this now dominant set of social relations in the old industrial societies. Looked at from right outside, the traffic flows and their regulations are clearly a social order of a determined kind, yet what is experienced inside them ... is movement, choice of direction, the pursuit of self-determined private purposes. All the other shells are moving, in comparable ways but for their own private ends. They are not so much other people, in any full sense, but other units which signal and are signalled to, so that private mobilities can proceed safely and relatively unhindered. And if all that is seen from outside as in deep ways determined, or in some sweeping glance as dehumanised, that is not at all how it feels inside the shell, with people you want to be with, going where you want to go. Thus at a now dominant level of social relations, systems quite other than settlement, or in any of the old senses community, are both actively and continually reproduced.

The car, in short, privatises the public sphere of the road and the street and encloses the global commons of air, land and water for private use. Unlike the bourgeois public sphere of early modern Europe idealised by Habermas in the coffee house and periodical press in which issues of common concern could be debated (see Giblett, 1997b and 2004, chapter 2), in hypermodernity the petit bourgeois private sphere of the car and the petit bourgeois public sphere of the road attenuate communication. On the open road in the country the eye is free to travel over the panoramic landscape laid out before it, while the body as a whole is imprisoned in a metal box. 'The urban motorway', for Virilio (2005b, p. 59), is 'the concentration camp of speed' entailing 'segregation and incarceration.' Cars are also beautiful, if not sublime, objects. Cars, for Benjamin (1979, p. 230; 1999b, p. 211), are 'as beautiful as armour from the age of chivalry'. Cars entrap and empower, they are a suit to wear and a weapon of war. They are made for protection from the world outside and assault against it. They combine aesthetic pleasure and mechanical function.

Yet the aesthetic pleasure the car proffers is limited, not only to the sense of sight but also in range. For Friedberg (1993, p. xii), 'the private mobility of driving transforms the windshield into a synoptic view' rather than into a panoptic view. Drivers communicate with other drivers by the limited language of horn and lights, or their car's occupation of space (often in positions of aggression), or by gestures in a poverty-stricken code characterised largely by those of abuse and occasionally ameliorated by those of politeness or apology. Larger issues of public concern about the car such as deaths and injuries it causes, the location and building of highways, bridges, freeways and tunnels, and its eco-impacts of pollution and global warming can hardly be discussed in this context! The car operates in the public sphere, but it is not the Habermasian public sphere where issues of common concern can be discussed.

Besides an instance of 'mobile privatisation' the car is equally an instance of private mobilisation in which the private sphere is mobilised against the public sphere of the street and the road. The car is also mobilised against the global commons of the biosphere, the atmosphere, the hydrosphere and the lithosphere, what I call collectively the '-ospheres' (see Giblett, 2004, chapter 2). Whether in the private sphere of the car, the office, or the home in the suburbs, or in the public sphere of the street or road, the modern is always already in the -ospheres, the global commons of air, soil, water and the local bioregions of watersheds. All these spheres are encompassed in the ecosphere of the earth household.

The car mobilises private individuals for the army of late modernity in the war against 'nature', or the -ospheres, at the same time as it is what Cubitt (2000, p. 140) calls 'a device for immobilisation and subjection'. The car mobilises the private sphere of interior space against the public sphere of the street and the road, and against the -ospheres of 'nature', at the same time as it immobilises the body of the driver and passenger.

The car of neo-colonial hypermodernity has replaced the sheep and cattle of colonial modernity as the foot soldiers in the war against nature (see Giblett, 2004). The car privatises, encloses and consumes the public sphere of the street and the road and the cultural landscapes of city, suburb and settled country as well to some extent the indigenous cultural landscape of Aboriginal country or 'wilderness', and the global commons of the -ospheres. The car consumes greedily cultural and natural landscapes, human lives (25–30 million so far) and other non-human lives. For some early writers on the car (cited in Sachs, 1992, pp. 6, 14 and 26), including the Decadent Alfred Jarry, more recently for the poet W. H. Auden (cited in Williams, 1991, p. 86) and even more recently for biologist David Bellamy (1998), the car is a monster. In less emotive terms, for Cubitt (2000, p. 139), 'the private automobile is without doubt the single most deleterious innovation of the last century'.

These critics of the car are unlikely bedfellows. Alfred Jarry's (1999, pp. 25, 42 and 47) novel *The supermale* of 1902 includes 'a monstrous automobile...propelled by explosive mixtures'. The car is described as 'the metallic beast, like a huge beetle'. In a bizarre Kafkaesque metamorphosis, instead of, as in Kafka's story, Gregor being transformed into a monstrous dung beetle, in Jarry's novel the beetle has been transformed into a monstrous car. For McLuhan (1964, pp. 224 and 225) too 'the car has become the carapace, the protective and aggressive shell, of urban and suburban man'. The car is a monstrous beetle (and not just a Volkswagen). In a counter-metamorphosis, the car is also a monstrous mechanical dung beetle that transforms living beings into excrement. The car in Jarry's novel is also described as 'a fantastic machine, scarlet and snorting' which 'exhibited without modesty, almost with pride, its organs of propulsion. It seemed like a lewd and fabulous god carrying off the girl' driver. The car was to be apotheosised, albeit uncritically, in similar terms a few years later by the Italian futurists in their First Manifesto of 1909, a hymn to 'the beauty of speed' encapsulated in 'a racing car whose hood is adorned by great pipes, like serpents of explosive breath' (cited in Hughes, 1980, p. 43).

As both magician's bower and monstrous machine the car encapsulates the ambiguity and paradox of modernity, its defining feature for some. 'To be modern', Berman (1983, p. 15) argues, 'is to find ourselves in an environment that promises us adventure, power, joy, growth, transformation of ourselves and the world – and, at the same time, that threatens to destroy everything we have, everything we know, everything we are.' Modernity is fissured and divided against itself by the promise of power and the threat of destruction. This ambiguity is produced by the modern industrial technologies that both promise power and deliver destruction. The car is no exception to this general rule. Indeed, it is perhaps a truism, if not a cliché, to call the car, as Sachs (1992, p. 32; see also pp. 35 and 58) does, 'the symbol of modernity' – not least because it both promises power and brings destruction. Beneath the surface of its gleaming exteriors, the depths of its throbbing engine represent paradoxically speed and power, mayhem and destruction, death and injury, pollution and global warming.

Like the railway, telegraphy, photography and the cinema, the car is a product of the masculinist project of modernity to transcend space and time. In 1906 the promise was made that 'the automobile ... will grant human beings their conquest over time and space by virtue of its speed of forward motion' (cited in Sachs, 1992, p. 9). Speed conquers time by conquering space. The quicker it takes to travel between two places, the less far apart they seem to be. But the car, like all modern industrial technologies, never delivers unequivocally on its promise. The speed of forward motion is not so much dependent on what is beneath or around one, what sort of car one is driving, but on what is before one, what is in front of one's car, whether it is an open road or a traffic jam in the best and worst case scenarios. The car promises conquest over time and space but imprisons the driver in the suspended time of sitting in the enclosed space of one's car.

The early utopian period of the uptake of any new technology is revealing as we have seen on many occasions previously. In 1906, inflated promises and wild claims were made that

> the automobile has succeeded in overcoming space and time more completely than ever before. From a certain point of view, all of the technological endeavours of mankind are directed only at transcending space and time. The telegraph transmits the written word over great distances in the smallest time units, just as the telephone does for the spoken word ... Steam locomotives, and electric trains, steamships, bicycles, airships – all of these serve the purpose of

conquering space and conquering time, to race across the greatest possible distances in the shortest possible time.

(cited in Sachs, 1992, p. 167)

This writer ultimately sees no distinction between modern communication and transportation technologies; both are in the business of transcending space and time.

Yet communication technologies take off, and take over, from the point where transportation technologies run up against the limits of their ability to conquer space. The instantaneity of communication technologies overcomes the duration of transportation technologies. 'To annihilate spatial distances by means of mechanical force,' according to Sachs (1992, p. 215),

> that is the motivation that has driven energy-intensive transportation from the first locomotive to the space shuttle. And it has now run up against its limits. A new cultural motivation has entered the scene with communications technology: to annihilate spatial distances by means of processing intelligence. Its attractiveness derives from adventures in mental space rather than adventures in geographical space. Now that the conquest of geographical space ends as often as not in traffic jams, the home computer comes along just in time to reinvigorate the enthusiasm automobile drivers have lost.

And so the Internet comes along just in time to provide an information superhighway on which to drive one's personal computer and conquer time and space just when one's car is increasingly snarled in traffic jams and fails to deliver on its promise of the conquest of time and space. But then, of course, the Internet becomes caught in its own traffic jams of spamming, protracted download times and viruses.

The home computer is to the car on the suburban street and the country road as the Internet-capable computer on the information superhighway is to the car on the intercity freeway. Sachs (1992, p. viii) asks rhetorically, 'is the microchip not for our children what the engine was for our grandfathers?' Later he answers his question explicitly by suggesting that 'the current enthusiasm for microcomputers...reflects a leitmotif or two from the early years of the automobile' (Sachs, 1992, p. 175). The shift from transportation to communication as the leading edge technology is a transcendence of the geographical into the virtual. The shift from the macadam highway to the information superhighway,

from views from the road to views on the monitor, from the wind-screen to Windows®, from the highway network to the Internet, from transportation to communication represents not only the 'progress' of modernity but the shift in the conquest of space and time from trans-portation technologies to communication technologies.

What Sachs (1992, p. 213) calls

> the sign of superindustrialization is not the unconcerned plundering of natural resources of old, but has as its core the project of getting as much as possible out of limited resources. It aims at draining the flood of environmental burdens by means of countless canals, hoping in the process to rescue desire. The computerization of society, ultimately a technological-cultural salvage project, is presented as the sanitiser of the old industrialization.

Just when the old industrialisation shows itself to be dirty and greedy, the new industrialisation comes along all squeaky clean, not only to promise the redemption and sanitisation of the old industrialisation, but a whole new bright and shiny future of its own.

The magic of the car not only resides in stasis in its sweeping fender and swelling hood, nor only in movement in its power and speed, but also in its private domestic interiors. Virilio (2000b, p. 25; see also 1991a, pp. 78–79 and 2005b, p. 55) traces how 'the private car was originally part of the furniture boom' and how it 'actually appeared as the inheritor of the seat that *rests the lower limbs*' so that it is in fact 'a *moving piece of furniture* . . . if not a prosthesis for the physically disabled'. The car not only enables the human body to move from place to place, but also disables it/us by making it/us sit in a chair for however long the journey takes.

If, as Virilio (1989b, p. 31) argued elsewhere, cinema is the public secular cathedral in which black masses were performed, then the car is its private counterpart. Cinema is Roman Catholicism to the car's Prot-estantism. With the car, like the printed vernacular Bible, every man is his own priest, performing his own communion in private and inter-preting the text of the world as he or she likes. The catchcry of the religion of modernity is 'progress' and its material expression is in what Sachs (1992, p. 219) calls 'speed machines'. The car enacts the religion of progress that valorises speed as mystical ecstasy and fetishises the car as the elements of bread and wine in an unholy mass to be transubstan-tiated into an experience of pure speed and consuming greed.

The car is not only the symbol of modernity, of modern industrial capitalism and urbanisation, of power and freedom; it is also
the symbol of modernism, of a cultural aesthetic in the service of
modernity. Cars, Pickett (1998, p. 11) argues, 'from the start, were bearers
of modernity and modernism ... cars embody a distinctive twentieth
century aesthetic'. This combination of utility and aesthetics, of industry
and design, prompted Barthes (1973, p. 88) to call the car 'the Gothic
cathedral of modern times':

> the supreme creation of an era, conceived with passion by unknown
> artists, and consumed in image if not in usage by a whole population
> which appropriates them as a purely magical object.

Unlike the public sphere of the Gothic and cinema cathedrals, the car
is a private space of magical domesticity.

The car, for Barthes (1973, p. 88), is the modern equivalent of a Gothic
cathedral not only because it is consumed by users who have little idea
of how it is built or how it works but also because it is a transcendent,
magical object:

> It is obvious that the new Citroën has fallen from the sky inasmuch
> as it appears at first sight as a superlative *object*. We must not forget
> that an object is the best messenger of a world *above* that of nature
> [as the sublime was seen by Kant as the faculty by which we regard
> ourselves as independent of nature (see Giblett, 1996, chapter 2)]:
> one can easily see in an object at once a perfection and an absence
> of origin, a closure and a brilliance, *a transformation of life into matter*
> (matter is much more magical than life), in a word a *silence* which
> belongs to the realm of fairy-tales [my emphasis].

The car is both a sublime communication technology as it transforms
base matter and the space and time co-ordinates of everyday, pedestrian
life into speed and transcendence of space and time, as well as the
sublimate of the commodity fetish that transforms that experience of
life on a higher, sublime plane back into dead matter.

Yet the sublimate of the commodity fetish is by no means mute as
it communicates 'news from nowhere', from utopia, that sublimated
world above nature, back to messy and organic everyday life. The slimy
and organic life of raw materials and human labour is sublimated into
the secular heaven of the sublime that transcends space and time only
in that process to be transformed into the sublimate of the commodity

made of dead matter. The magic of modernity appropriates pre-modern magic and creates a magical world *above* nature rather than respecting and conserving the magical world *of* nature. Serres (1982a, p. 110) argues that 'once the sacred is placed outside of the world in faraway locations which are of no interest to us, Nature is born, objectively, bearing her own laws'. Nature, in short, becomes Natural History (see Giblett, 2004, chapter 1).

The car not only produces an interior space but also entails a reconstruction of exterior space. One's windowed shell always gives a view of a landscape, a cultural construction or reconstruction of 'the environment'. For Sachs (1992, p. 189), 'the domination of distance has brought with it a revaluation of space: the gaze is directed at distant locales beyond the neighbourhood, and the immediate spatial world has declined in significance'. Life is elsewhere, to borrow Milan Kundera's phrase, the bane of modernity, rather than life is local, the conservationist's motto (Suzuki, 1998, p. 49). The car, for Virilio (1991b, p. 67), condensed 'without distinction the near and the far' in what he calls 'an abusive interlocking'.

The role of the car as vehicle, Virilio goes on to argue,

> consists less in transporting the passenger than in sliding aside physical reality, modifying first all lenses, then all optical instruments, and finally all different planes of visual experience ... The role of the means of communications moves from displacing the users to detaching them from their immediate environment.

The means of communication shifts from moving users from one place to another to removing users from their place, from their sense of place.

The car inherited the mobile view from the railway. For Sachs (1992, pp. 154 and 5), 'seen through a train window, the landscape offered itself up like a panorama of ever-changing scenes ... The railway choreographed a landscape that then took on a monumental dignity.' The rigid lines of locomotion and the fixed scheduling of stops dictated an immobile viewer, panoramic views and a largely static tableau of monumental landscapes, such as the early national parks (see Giblett, 2004, pp. 145–146). The flexibility of the car's journey, however, including stoppages made from personal preference, produced a more dynamic taking of views and a more mobile viewer. Indeed, the car transformed the armchair tourist of the railway passenger into the ersatz explorer of the driver. For Sachs (1992, p. 155), 'the automobile broke the rigid perspective as seen through the train window ... the automobile transformed the traveller into a potential explorer...now, with active probing, the tourist's gaze could penetrate even the most distant corners'. The car penetrated space and enabled the tourist's gaze to penetrate place. The tourist could go off the beaten track to explore local back-roads and gawk at backward local people, their houses and customs.

The car, Sachs (1992, pp. 45 and 49) argues, 'comes into its own only when space is penetrable and speed unimpaired...A totally new type of road was required to make space penetrable.' The penetrability of space and place for the tourist by means of the car occurred via the national highway on which the driver as private individual drove. The railway had a nationalist impulse that appealed to public-minded citizens, whereas the car had an individualist impulse appealing to private consumers. The railway, according to Sachs (1992, p. 104), 'enticed travellers to regard themselves as masters of the nation (and later, the airplane as masters of the world)'. The car enticed drivers to regard themselves as masters of the city and the country, the dyadic landscapes of modernity that the vectors of the car conjoin and destroy. The car, for McLuhan (1964, p. 224),

> ended the countryside and substituted a new landscape in which the car was a sort of steeplechaser. At the same time, the motor destroyed the city as a casual environment in which families could be reared.

The car brought the city and the country closer and destroyed the distance between them, but at the same time destroyed them and the distinction between them by making the landscape (urban and rural) into a medium for cars. Everywhere became car-city, or car-ntry. The car and other means of communication altered perceptions of space. The reduction of distances, for Virilio (1986b, p. 133),

> corresponds to the negation of space...*In fact, the strategic value of the non-place of speed has definitively supplanted that place.*

Speed killed place. Here today, gone tomorrow.

The car and other means of communication altered perceptions not only of space but also of time. Everywhere became car-city, or car-ntry, not just spatially but also temporally. For Virilio (1991b, p. 95),

> The rule of space and the spatiality of different rural and urban places gives way to temporality, and the nature of regimes of temporalities, produced by the advanced technologies.

Time rules, okay? No longer located primarily in a spatial grid of longitude and latitude distanced from the city, the local location is positioned in a temporal zone the same as, or distinct from, metropolitan centres and in near-instantaneous communication with them that transcends distance. The distance between two points counts less than the time taken to travel between them, whether it is via transportation or communication. The faster the speed, the less time expended in getting from A to B. Places become what Virilio (2005b, p. 50) calls 'points of arrival and departure'. Goegraphic place, for Virilio (2005b, p. 77), is 'no longer the foundation of human experience but rather a pole to be reached in the round-trip exercise'. The result, for Virilio (2005b, p. 158), is that 'the benefit of the time saved must be measured against the loss of value of the space passed through. The rapidity of automobile trajectories only increases at the cost of the meaning of places.' The peaceful identities of places are obliterated by the speed with which the car driver and passenger slip through them on the way from here to there.

A straight line is the shortest distance between two points, whereas speed, for Virilio (1991b, p. 117), is 'now the shortest route between two points'. Speed, for Virilio (1986b, p. 22), is also 'time saved'. As speed, for Sun Tzu (quoted by Virilio, 1986b, p. 133; 2005b, p. 102), is 'the essence of war', saving time is waging war. Speed is war (see Virilio, 2005b, pp. 59 and 65), or more precisely what Virilio (2005b, p. 117) calls 'pure war', or what he (1986b, p. 139) calls the 'last war', the war against time, and space, against nature. Speed militarises civilian life. What Virilio (1990, p. 92) calls 'the saturation of time and space by speed mak[es] daily life the last theatre of operations, the ultimate scene of strategic foresight'.

The car transformed the landscape, especially by imbuing the national highway with national significance. In a tellingly entitled book, *The Automobile Conquers the World*, published in 1938, the writer claimed that 'the national highways ... [are] the noblest adornment of the noble German landscape ... [and] the supreme crowning of the landscape' (cited in Sachs 1992, p. 52). German National Socialism sublimated the landscape of trench warfare of the First World War into the landscape of the national highway. This sublimation was a way of forgetting the defeats of the immediate past and its blighted landscape, of linking up with a rosier romanticised past of postulated closeness to nature and of forging a utopian way into a techno-pastoral future. The national highway was a way of having one's nostalgic cake and eating one's utopian cake too.

In the late nineteenth and early twentieth centuries the nation had sought its identity in the national landscapes of the national park to which the citizen was brought initially by the railway and the consumer later in larger numbers by car. National parks were nationalist (see Giblett, 2004, chapter 8). The nation-state had also constructed its identity by defining its boundaries against the boundaries of other nations in the case of the settled colonisers, or its frontier against the wilderness in the case of the settler colonised (see Giblett, 2004, chapter 6). National highways were nationalist too. The national landscape of the twentieth century was the intra-national highway that linked the nation together by internal routes of communication and communicated to its citizens pleasing prospects of monumental landscapes imbued with national significance, or of retrospective rural landscapes endowed with nostalgia going back into the past, or of prospective utopian vistas leading on into the future.

Nazi German highways were supposed to be, Sachs (1992, p. 55) argues,

> monuments for the future, pyramids for the Reich – this is what freeways were supposed to be, not soulless strips of concrete. The much-sung German landscape was to be enhanced, not destroyed. What would dominate was not the straight line, but the gentle curve; not the throughway, but the beautiful view.

The German feeling for landscape, that Benjamin saw had had its inevitable outcome in the landscape of trench warfare in the First World War, was revived in the national highways of the Nazi prelude to the Second World War. The national highways were monuments to German National Socialism just as the national parks were monuments to American nationalism in the nineteenth century (see Giblett, 2004, chapter 8) and as national highways became in the twentieth. Both subscribed to a monumentalist landscape aesthetic with a touch of a nostalgic and sentimental pastoro-technical idyll.

Nazism was a heady and contradictory brew of modernity and nostalgia, culture and nature, technology and ecology. The national car and the national highway were sites and symbols of this contradiction. Nazi propaganda, for Cornwell (2003, p. 9),

> Gave an impression of technocratic modernity matched with a restored and bracing naturalism – exercise, healthy living and outdoor pursuits. The making of the Volkswagen, the people's car, and the

great autobahns (known as 'Adolf Hitler's roads') were symbols of a modernizing nation state, signalling in the propaganda of the time, a marriage between motorized transport for all and the beauty and freedom of the landscapes of the Fatherland.

Nazi technocracy, Cornwell (2003, p. 9) goes on to argue in relation to what he calls these 'landscaped highways', was 'seldom unmixed with mawkishly bucolic dimensions'.

Unlike the rectolinear grid-plan town that subjected the colonial landscape of the nineteenth century to a rationalist logic and a militarist imperative (see Giblett, 1996, chapter 3), the gentle, sweeping curves of the national highway of the twentieth century would subject the landscape to the aesthetics of the beautiful and the picturesque. Both German National Socialism and American nationalism imbibed and reproduced this modern landscape aesthetic in their national highways. Rather than the whipping boy of the left or the bad dream of the right that became a reality, Fascism was modernity and modernism taken to the nth degree (see Giblett, 1996). Yet Fascism is not just a thing of the past but is still with us to the extent that modernity and modernism are taken to the nth degree. And to the extent that Fascism is inscribed in and on the human body. Adorno argues that 'the movements machines demand of their users already have the violent, hard-hitting, unresisting jerkiness of Fascist maltreatment' (cited in Williams, 1991, p. 90). Fascism is inscribed or programmed into the movements of the driver, the car cyborg.

Modernity and modernism enacted a class politics over landscape by the use of industrial technology in which the bourgeoisie took over the superior position of the aristocracy. The car was one vehicle for doing so. In the beginning of the car industry, as Sachs (1992, p. 36) puts it, was 'the luxury automobile' and 'the luxury market, rather than mass consumption, was the midwife of the automobile'. Luxury, that hallmark of private consumption, rather than speed, and certainly not efficiency, was the midwife that assisted at the Bachelor Birth of the Bachelor Machine of the car. Luxury was a class privilege and the car assured its class patrilineal descent. The aristocracy and the bourgeoisie used, as Sachs (1992, p. 10) puts it, the automobile's 'speed and power to display their social superiority. It had always been a sign of power to master an extended spatial range and leave others staring in one's wake.'

Mastery of time and space was strongly tied to the desire for mastery of the social. Sachs (1992, p. 12) goes on to argue that 'the masters of time and space demanded that they also be the new masters of the social order. And, reversing, the logic, the new masters of the social order documented their claim through their new power over time and space.' Yet, as Sachs (1992, p. 187) goes on to point out, 'the masters of space and time awaken to find themselves slaves of distance and haste'. The car driver dreams of mastery of space and time, but wakes up to find himself a slave of a point in space and time, and to a schedule to be here now, and then there. The driver is handcuffed to the steering wheel with a wrist-watch. The car not only gives the driver power over other classes and the landscape, but also enacts power over the driver. There is no instrument for wielding power that does not also wield power over the wielder.

Greed for speed

Driving a car fast produces in drivers what early writers on the car described as 'the feeling of rising and flying "beyond limitations"', 'a mix of anxiety and pleasure' in which the driver 'rise[s] *above* nature' (Sachs, 1992, pp. 112 and 113 my emphasis). The car is a sublime technology that encapsulates the capacity by which, as Kant put it in a discussion of the sublime, we calculate ourselves as independent of nature (see Giblett, 1996, chapter 2). Transcendence of nature includes transcending our own bodies through what Sachs (1992, p. 3) calls 'the liberation of speed from the fetters of corporeal nature'. Yet this liberation comes at the cost of imprisoning corporeal nature in a speeding metal box.

The mix of anxiety and pleasure that the car produces is a sublime aesthetic experience. For Burke and Kant, the sublime produces a pleasure bordering on pain, an experience tinged with terror associated with the phallus, or law of the father (see Giblett 1996, chapter 2). The car is a phallic instrument that enacts a sublime masculinist politics and poetics of space and the landscape (see Giblett, 1996). In the secular theology of the sublime, speed is the sacrament and the car is the vehicle for apotheosis. Speed, for the Italian Futurist Marinetti (Flint, ed., 1972, pp. 95, 96) writing in 1916, 'finally gives to human life one of the characteristics of divinity: *the straight line* ... The intoxication of great speed in cars is nothing but the joy of feeling oneself fused with the only *divinity*', the divinity of speed. The car, Patrick White claimed, 'made God unnecessary. Speed, after reducing your flesh, leaves you on equal terms with the natural forces that have replaced Him' (cited in Pickett, 1998, p. 23). Or more precisely, the sublime forces of speed and power that drive the car have replaced God, rather than replacing the natural forces that 'through the green leaf drives the flower', as Dylan Thomas put it.

Besides subjecting the driver and passenger to the terrors of the sublime and of a theology without God, the car also encapsulates them in a cosy interior. For Sachs (1992, p. 131),

> vehicles become the repository of their drivers' desires for comfort, which appear to stem equally from simultaneous wishes for regression and omnipotence. One sinks in so cosily, almost as if into a womb, surrounded by humming technology and unmistakable comfort, while at the same time indulging a feeling of power and force rarely experience in life otherwise... It is both mysterious and familiar to sit in the snug warmth of the car... My car is my castle.

My car is also a lethal weapon. It is not mysterious and familiar to sit in the cold and twisted metal and shattered glass of a car crash. The car represents both the terrific (even terrifying) mystery of sublime speed and the horrific and horrifying mystery and unfamiliarity of the uncanny car crash that so fascinates Vaughan in Ballard's novel *Crash*.

The car as both sublime phallus and slimy uterus enacts a masculinist poetics and politics. For the masculinist driver the car is both sublime phallus and uncanny uterus. The car, as Heathcote Williams (1991, p. 66) puts it, is 'A pumping penile womb / With illuminated breasts, / And auto-erotic fuel injection / To achieve orgasmic speeds...' In Ballard's 1973 novel *Crash* the interior space of a car is feminised as a virginal body that procreates by immaculate conception. In other words, the car is a Bachelor Machine for a Bachelor Birth. There are repeated references in the novel to the car's 'immaculate body' (p. 163), 'immaculate vehicles' and 'immaculate machines' (p. 174). 'Immaculate' is a stock-in-trade term in the rhetoric of classified advertising for cars. Cars are what McLuhan (1967, p. 217; 1967 pp. 100–101) calls 'mechanical brides', womb-like interiors offering control of exteriors styled as love machines. A masculinist womb envy and desire for mastery gives birth to cars that kill.

Cars phantasise or project a fecund space but produce what *Crash* calls a 'metallised Elysium' (Ballard, 1995, p. 198). As Elysium is the home of the blessed *after* death, the driver is dead (or deadened) as soon as he or she steps into the dead matter of the car. 'In a car', Virilio and Lotringer (2002, p. 38) says, 'the body is dead.' The driver and passenger is the dead living, like the star and the fan, as well as driving dead. The passenger, for Virilio (2005b, p. 44), is 'nothing but a dead man who rides'. When the metallised Elysium drives in the 'Elysiums for gentlemen' (see Carter, 1989, pp. 202–229) of the grid-plan town, death is within and without and all around. The car-city is death-city, necropolis. The driver has made a Faustian pact with the devil to gain mastery over space and time in exchange for his or her soul, and body. Sitting in a car, the narrator of *Crash* describes how 'around me the interior of the car glowed like a magician's bower' (p. 199). If the car is a bower it is what *Crash* also calls a 'metal bower' (p. 202) of cold hard steel, a lethal weapon that not only kills others but kills its own driver with the death of a thousand cuts to his or her physical capacities.

The car not only supplements the lack of human capacity for speed but also atrophies the development of human capacities to perform other activities. The car is a prosthesis as it does what humans *cannot* do, rather than developing what humans *can* do. The car gave physical and psychological power to its driver. The car, Sachs (1992, p. 114) argues,

> in particular promised the feeling of omnipotence; of the tools that industrial progress has brought within reach of individuals, no other represented so disproportionate a gathering of impersonal energy to strengthen weak human powers ... an ego prosthesis.

And a leg prosthesis, a prosthesis for walking, both a vehicle for boosting one's ego and one's physical capacities. Sachs (1992, p. 135) goes on to argue that 'the essence of the comfort technology [is] that human senses, capacities, and responsibilities lay fallow'. The in-*car*-cerated human body is rendered largely passive and its capacities constructed as lack *supplemented* by a prosthesis, rather than an active body *complemented* by a tool or utensil.

Yet there is a difference between the driver and the passenger in the degree of activity and passivity imbued by the car. The driver inside the cosy womb of the car controls the body of the mother machine. The driver is no passive passenger carried along in the womb but the active controller of the car body. The car for the driver is not only prosthesis (supplement to lack) but also extension (complement to status quo). For McLuhan (1964, p. 221), 'the simple and obvious fact about the car is that, more than any horse, it is an extension of man [sic] that turns the rider into a superman. It is a hot, explosive medium of social communication.' The internal combustion engine is a contained explosion that, like the steam engine, transforms solid matter into energy and transports the occupant as bearer of information.

The car also marks the occupant as bearer of class, or at least of its trappings. Like Benjamin, for whom the car is armour, McLuhan (1964, p. 223) goes on to argue that 'the car gave to the democratic cavalier his horse and armour and haughty insolence in one package, trans-mogrifying the knight into a misguided missile'. And so into a lethal weapon. Just as the railway was the 'iron horse' the car was the steel horse, both of which are sublimated animals. The car elevated 'democratic man' up the social scale and ostensibly catapulted him into a new class by acquiring the trappings of power and prestige that had formerly been the exclusive privilege of the higher classes, both aristocratic and bourgeois. In the internal combustion engine, for Fuller (1961/1992, p. 137), 'lay hidden the greatest revolution in civil life and war since primitive man [*sic*] first tamed the horse'. The car is the steel horse, the direct descendant of the iron horse of the railway with similar class trappings.

The car not only fills up limited human capacities for mobility in terms of range and rapidity, but also adds to those capacities by providing an extension to those capacities in terms of power and speed. Drivers, as Sachs (1992, p. 133) puts it,

> must extend their region of sensory attention beyond themselves to the whole vehicle and synchronise its mechanical functions with bodily movements, indeed, even internalise the ensemble of 'man and machine.' ... It is cosy, this tension between security and the thrill of power, between uterus and phallus.

The driver becomes a cyborg, no longer just an organic being but an amalgam of biology and machine, of living and dead matter.

The combination of prosthesis with power over space and time, of cosy womblike interior and destructive phallic exterior produces a greedy oral-sadistic monster that devours human and other lives, eats up land-scapes and bioregions, consumes resources, excretes polluting wastes, warms the globe, upsets weather patterns, changes climate, raises sea levels and so on. Early writers on the car described it as 'the motor-ised monster' and as 'the gasoline-driven monster' (cited in Sachs, 1992, pp. 6, 14 and 26). As W. H. Auden put it,

> Dark was that day when Diesel
> conceived his grim engine that
> begot you, vile invention ...
> metallic monstrosity,

bale and bane of our Culture,
chief woe of our Commonweal.

(cited in Williams, 1991, p. 86)

This monster has an insatiable appetite for consuming distance and is a prosthesis not only for walking but also for the gaze. 'The automobile's greed for distance' is the mechanical and commodified satisfaction of 'the gaze greedy for distance'. This greed to the power of two gave rise to 'highways eating up the landscape' (Sachs, 1992, pp. 174, 189 and 190).

The car greedy for distance, the driver's greed for speed and his or her gaze greedy for distant places and the highway eating up the land-scape produce an oral-sadistic monster that consumes gluttonously and greedily, not gratefully and generously (see Giblett, 2004, chapter 9). The car is a monstrous machine that consumes landscapes, drivers, passen-gers and pedestrians human and non-human. The petroleum-powered car inherited the monstrous, oral sadistic features of the older modern technologies of the steam-driven, coal-powered phase of industrial capit-alism. These included the steam trains that ate up the bush, the steam dredgers that devoured marshes and other wetlands (see Giblett, 1996), the steam saws that chewed up forests and the crushing machines that ate up rock (see Giblett, 2004, pp. 115–116 and chapter 9). The car consumes the bush, the wetland and the forest and creates national landscapes of sweeping vistas, but bequeaths a diminished urban and rural landscape, and even a dead environment, to its children. Unlike the phantastic monsters of myth and legend killed by human heroes, the monster that is killing us is us; the monster is us.

7
The Magical in The Modern: Ethereal Radio

Radio not only arose out of telegraphy, as Williams (1989, pp. 120 and 121) points out, but also, according to him, was 'at first conceived...as an advanced form of telegraphy' (Williams, 1974, p. 12). Radio communicated over long distances like telegraphy, but it made improvements over its predecessor by freeing communication from a vulnerable physical connection (such as terrestrial wire or submarine cable) between two or more places. Early radio was 'wireless' (Tucker, 1978, pp. 1230–1244; Early, 1996, p. 325). Early wireless, according to McLuhan (1964, p. 304), was 'regarded as a form of telegraph'. Telegraphy separated communication from transportation for the first time; radio went one step further by separating communication from a direct physical connection between sender and receiver and by creating a metaphysical connection between them.

Radio was not only wire-less communication unlike telegraphy, but also sound transmission and communication like telephony. Communicating sound electrically without using wires was a development of telephony and telegraphy. Radio, for Hanson (1982, p. 27), 'was an attempt to combine the best of both worlds – electrical voice and sound communication without wires'. To do so radio used the resources of the electromagnetic spectrum (see Lax, 1997, pp. 8 and 26–28). This consists, as Smythe (1981, p. 302) puts it, of 'the fields of magnetic and electrical force capable of transmitting electromagnetic energy in successive waves of different lengths'. As the first technological use of spectrum, radio was a radical departure from its predecessor in telegraphy and photography, both of which relied on communication by visible means whereas radio communicated by invisible means.

Yet radio was also a direct descendant of photography. If photography was literally 'writing with light' the shadows of objects in space, then with radio 'in the very air one breathes are words written by electricity' (cited in Kern, 1983, p. 67). Yet radio was invisible writing, unlike photography (and telegraphy). Radio was a kind of writing in invisible or inaudible ink that became audible when transmitted to, and decoded by, a receiver. Radio inscribed and colonised invisible space. Radio wrote words in what we now call 'electromagnetic spectrum' but what was called 'the ether'. This was the means to communicate without a direct physical connection between sender and receiver.

Ether

By transcending the physical in ethereal realms, radio was what could be called 'the first metaphysical means of communication'. It prefigured subsequent metaphysical technologies of the twentieth century. Technology for Ernst Jünger writing in the 1920s is 'the real metaphysics of the twentieth century' (cited in Spretnak, 1997, p. 128). Modern communication technologies are metaphysics in action. In the late nineteenth century radio put metaphysics into action and ushered in the twentieth-century triumph of physics over the physical, and the twentieth-century metaphysics of communication technologies. Radio is comprised of physical components of the transmitter and the receiver. Between them lies the invisible medium of the electromagnetosphere, or electromagnetic spectrum, part of what used to be called 'the ether'.

The 1920s, according to Miller (1997, p. 49), was the era of 'faith in the "ether", a semi-metaphysical concept that was used to describe the air through which radio waves passed'. Yet the medium through which radio waves pass is not only, or just, air in the strict sense of a mixture of oxygen and nitrogen. The medium is the electromagnetosphere, not limited or restricted to the atmosphere, but also including the ionosphere. The atmosphere and the electromagnetosphere are not co-extensive; the electromagnetosphere is more extensive than the atmosphere, though both are terrestrial phenomena. The ether not only transcends the earthly into the extraterrestrial, but is also a terrestrial product of the earth's electromagnetism.

The 1920s may have been the heyday of 'faith in the ether', but belief in the ether predates that decade by two millennia or more. Moderns exercised faith in the ether, but the ancients propounded the philosophy of the ether. For the pre-Socratic philosopher Heraclitus the 'aither' was 'the brilliant fiery stuff which fills the shining sky and surrounds the world ... widely regarded both as divine and as a place of souls' (Kirk *et al.*, 1983, p. 198). It is 'the purer upper stratum of air (approximately the stratrosphere)' (Hammond and Scullard, 1970, p. 33). It contrasts with, and is opposed metaphysically and spatially to, the grotesque lower stratum of the earth and its depths. It is the heavens above the earth below, and above the depths below the earth.

In the Platonic dialogue *Phaedo* (Plato, 1969, p. 109a) Socrates defines 'the Ether' as 'the starry heavens' in which the earth lies. Both heavens and earth are pure, but water, mist and air are the dregs of the ether. These dregs drain into the hollow places of the earth where we live in 'measureless mud and tracts of slime'. Socratic dualism poses a pure and ethereal heaven and earth to impure and slimy sloughs and swamps (see Giblett, 1996, chapter 2, especially p. 38). The former are valorised and the latter denigrated. Socrates concludes that amongst these things below 'nothing is in the least worthy to be judged beautiful by our standards. But the things above excel those of our world to a degree far greater still.' For Socrates, the ethereal and sublime world above is far superior to the slimy and swampy world below.

Likewise, for Plato's student Aristotle (1984, p. 451) in his treatise 'On the Heavens (270b20-24), 'the primary body is something else beyond earth, air, fire and water'. For Aristotle ether is the fifth element that makes the other four possible. In particular, *aither* makes possible 'the eternal circular movement of the heavenly bodies' (Lloyd and Sivin, 2002, pp. 149 and 170). For Aristotle (1984, p. 451), to 'the highest place the name of *aether,* derived from the fact that it "runs always" for an eternity of time', was given. The ether not only had a spatial location above and beyond earthly space but also had a temporal presence in and through earthly time. The ether is a space–time matrix and a medium of and for communication. It transcends local time and place and enables communication in real time through space. This is applicable not only to modern radio but also to hypermodern telecommunications.

By transcending time and space, and the earth below and the below-earth, the ether is also sublime. The ether, for Aristotle (1984, p. 627) in his treatise 'On the universe' (392a5-10), is 'the substance of the heaven and stars'. Stars are made of ether; the heavenly body of the star is ethereal. Aristotle goes on to state in the same treatise (392a29-30) that 'the limit of the ether ... encompasses the heavenly bodies'. The star has no existence outside the ethereal. The star transcends local time and place but at the cost of becoming ethereal. The star is located in divine realms above and beyond the earth. For Aristotle the ether is 'an element other than the four pure and divine', whereas earth, air, fire and water are impure and human.

The idea of the ether was taken up in England with the revival of interest in classical learning in the English Renaissance during the Elizabethan period in the late sixteenth century and Jacobean period in the early seventeenth century. Some Elizabethans were Aristotelians for whom the ether was, Tillyard (1943, p. 47) points out, 'a fifth element and the substance of all creation from the moon upward'. The ether is the fifth element that pervades or relates to the four elements of earth, air, fire and water (see Giblett, 1996, pp. 156–162): it transcends earth into air; it is the fire of lightning; and it flows like water, or at least 'electricity behaves like an incompressible fluid' (Smith-Rose, 1948, p. 28).

This apt description still persists. Electricity, for Taylor (1975, p. 11), is like 'some sort of invisible liquid' and like 'a strange ethereal gas' that 'does, at times, behave like a fluid'. At other times it performs what he calls 'ethereal dances' (Taylor, 1975, p. 180). Dancing ether is like, or even is the predecessor to, the youthful goddess of information performing her provocative belly dance in which she poses her heavenly body on the silver screen. The ether was feminised, fetishised and commodified just like information, advertising pictorialism and the cinematic star before her. She was made to dance and sing her siren song on the ethereal hearth of the home and in the chambers of the ear that addresses hearing, that most seductive of senses as it maintains and overcomes distance at the same time.

Returning to classical and Renaissance ideas, in the great chain of being the ether was the sublime upper realms up which 'man', as Raphael in Milton's *Paradise Lost* (V, 483, 4) puts it, 'by gradual scale sublim'd / To vital Spirits aspire'. By aspiring to transcend the earth into the sublime realms of the ether, 'contrariwise the earth itself was gross and heavy', as Tillyard (1943, p. 47) surmises, 'the cesspool of the universe, the repository of its grossest dregs', 'a place that hath received all the filthiness and purgings of all other worlds and ages.' The ether was at the opposite end of the scale to black noise and the grotesque lower earthly stratum (see Giblett, 1996, chapter 6). The ether was pure and light, the heavens of the universe, the summit of its highest sublimation, a place where all the refinements and sublimations of all other worlds and ages accumulate in glittering celestial array. The ether is 'the pure air surrounding the heavenly spheres', 'an air above the air' (Tillyard, 1943, pp. 51 and 53). The ether is valorised; the earth thereby denigrated.

The ether is the spiritual realm where, as in the opening lines of Milton's *Comus*,

Before the starry threshold of *Joves* Court
... those immortal shapes
Of bright aerial Spirits live inspheared
In regions mild of calm and serene Ayr
Above the smoak and stir of this spot,
Which men call Earth, and...
Strive to keep up a frail and feverish being.

Seventeenth-century anthropogenic air pollution in London is hardly any excuse for denigrating the earth and desiring to transcend it into the ethereal realms. The seventeenth-century Puritan equation of life with spirit is the driving force that denigrates life. Life, for Milton, was equated with breath and the spirit; death with living matter and slime (see Giblett, 1996, pp. 143–145). Tillyard (1943, p. 52) comments on this passage from Milton that 'the spirit is an intellectual being living above the sublunary realm, the terrestrial cesspool...., in the ether, in the region of the incorruptible spheres on the border's of God's court, the Empyrean'. The ethereal is extraterrestrial; the terrestrial excremental.

What was common to both planes or realms were effluvial flows. Matter that flowed invisibly in the extraterrestrial or visibly in the terrestrial was effluvial. Like Milton, Robert Boyle (1675/1966, pp. 345–350) writing in the seventeenth century summed up electricity behaving like a fluid in one word as 'effluvia'. Any substance – electromagnetic, gaseous or liquid – that behaved like a fluid was considered to be effluvial from at least the seventeenth to the twentieth century. The effluvial view of the circulation and flow of substances went hand in hand with the miasmatic theory of disease that contended with the contagion theory until the late nineteenth century, especially in relation to malaria (literally 'bad air'; see Giblett, 1996, pp. 122–123). What Boyle (1675/1966, p. 345) called

electrical attraction (which, you know, is generally listed among occult qualities)...may...be the effect of a material effluvium, issuing from, and returning to the electrical body...[rather than] the effect of naked and solitary quality, flowing immediately from a substantial form.

Electricity resides in the force-field between bodies rather than in bodies themselves.

As did the ether. According to Whittaker (1989, p. 5), Descartes in the seventeenth century propounded the view that the ether was 'a medium which, though imperceptible to the senses, is capable of transmitting force, and exerting effects on material bodies immersed in it'. Whittaker (1989, pp. 5–6) goes on to trace the etymology of the word which

> Had meant originally the blue sky or upper air (as distinguished from the lower air at the level of the earth), and had been borrowed from the Greek by Latin writers, from whom it passed into French and English in the Middle Ages. In ancient cosmology it was sometime used in the sense of that which occupied celestial regions.

In Elizabethan and Jacobean cosmology it was used in the same sense, but Descartes changed all that as he was, according to Whittaker (1989, p. 6), 'the first to bring the aether [*sic*] into science by postulating that it had mechanical properties'. Descartes, in fact, postulated that many things had mechanical properties, including the human body.

Despite Descartes, eighteenth- and nineteenth-century physicists were the inheritors and perpetrators of the sixteenth- and seventeenth-century world picture, including the occult view of electricity. In 1766 Franz Mesmer 'had propounded the widely accepted theory that a 'subtle fluid... pervades the universe, and associates all things together in mutual intercourse and harmony'. This fluid was regarded as related to magnetism' (Batchen, 1997, p. 153). In 1794 John Leslie (quoted in Batchen, 1997, p. 154) discussed ' "the phenomena of gravitation and magnetism" in relation to accepted theories of "electric fluid"'. Similar views were later expounded by Friedrich von Schelling (quoted in Batchen, 1997, p. 153) in 1799 'that magnetic, electrical, chemical, and finally even organic phenomena would be interwoven into one great association... [which] extends over the whole of nature'.

Although, as Czitrom argues (1982, p. 62; see also Sconce, 2000, p. 61), 'the notion of mysterious, all-pervasive ether later became discredited among scientists, it served [James Clerk] Maxwell as a convenient fiction to help explain the presence and behaviour of electromagnetic waves'. Maxwell, for Smith-Rose (1948, p. 32; see also Aitken, 1976, pp. 21–23), was 'the greatest theoretical physicist of the nineteenth century'. The ether was a means for Maxwell to propose 'a theory of the *electromagnetic field*' (Winston, 1998, p. 32). According to Norbert Wiener (1954/1989, p. 19), Maxwell regarded light as 'a form of electricity that could be reduced to the mechanics of a curious, rigid, but invisible medium known as the ether, which at the time, was supposed to permeate the atmosphere, interstellar space and all transparent materials'. The ether was invisible electric light. Its study was a branch of optics, the science of light, and its processes were mechanical.

The ether hypothesis, according to Czitrom (1982, p. 64), 'enjoyed a wide acceptance by scientists in the late nineteenth century'. In 1883 Oliver Lodge (1883, p. 330; see also Aitken, 1976, pp. 80–178), for instance, propounded the idea of 'one continuous substance filling all space: which can vibrate as light; which can be sheared into positive and negative electricity; which in whirls constitutes matter; and which transmits by continuity, and not by impact, every action and reaction of which matter is capable'. The ether was to nineteenth-century physicists what DNA was to late-twentieth-century biologists: the master code of matter, both of which, curiously, were figured as a spiral, whether that of whirls or double helix.

The 'ether', Batchen (1997, p. 154) concludes, was 'considered to be the vehicle for these [magnetic, electrical, chemical, and organic] forces'. It was the vehicle that transported the forces of invisible nature and made radio communication possible. The communication technology of radio arose out of and was dependent upon the transportation vehicle of the ether. Radio is yet another instance of a communication technology that arose out of a transportation technology, or at least out of a metaphor for it. The vehicle of the ether transported radio.

Rather than 'vehicle', 'medium' was the preferred term used to describe the ether in the nineteenth century. In 1865 Maxwell (quoted in Pocock, 1988, p. 70) saw the ether as a medium 'capable of being set in motion by electric currents and magnets'. Maxwell (cited in Winston, 1998, p. 32) concluded that 'there is an ethereal medium filling space and permeating space'. In 1888 George Fitzgerald picked up the term 'medium' (quoted in Pocock, 1988, p. 70). Before the media was the medium (see also Sconce, 2000, pp. 21–58). The association with the other sort of medium who or which communicates with the dead was a happy one. Radio, like photography, was not so much a means of communication with the dead as with the disembodied and unseen voice on the air waves, the dead living of the sender, and later the broadcast radio star. The media are the medium for communicating with the dead living.

Radio reinforced the link between science and spiritualism that had been forged with telegraphy. If Victorians in the mid-nineteenth century, according to Noakes (1999, p. 421), 'often found it hard to distinguish between telegraphy and spiritualism', the more so Victorians in the late nineteenth century found it impossible to distinguish between radio and spiritualism. Whereas telegraphy communicated using electricity along a wire, radio communicated using electricity invisibly in the ether. How spiritualist can you get?! Not only a shared belief in ether but also 'a common belief in spiritualism further unified the group of applied scientists' (Pocock, 1988, p. 72). For one nineteenth-century advocate the electric spark, 'shadowy, mysterious and impalpable…, seems to connect the spiritual and the material' (cited in Czitrom, 1982, p. 9). Spiritualism was, or held out the promise of, communication with the dead. William Thomson, 'the celebrated architect of the Atlantic telegraph', defined telegraphy drolly as 'the art of interchanging ideas by means of dead matter occupying space between two intelligent beings' (Noakes, 1999, p. 422). Radio can be defined in analogous terms as the art of interchanging ideas by means of ethereal matter occupying space between dead living things.

The ether gave new meaning to Nicholas of Cusa's vision of 'a universe whose centre is everywhere and whose circumference is nowhere' and to Pascal's definition of 'Nature' as 'an infinite sphere whose center is everywhere and circumference nowhere'. The ether is the stuff of nature. Lévy (2001, p. 26) has recently invoked this definition in his projection that 'eventually there will only be a single computer' whose 'centre will be everywhere, and its circumference nowhere'. Yet networked computers do not have a centre, and rather than their circumference being nowhere, they will be everywhere. Every real city will ultimately, for Virilio (1997b, p. 74; 2005b, p. 92), be merely a suburb of the virtual world-city hypernetworked by telecommunications whose *'centre will be nowhere and circumference everywhere'*.

The ether, for Lodge (1883, p. 330), 'exist[s] equally everywhere'. Like Nature, the ether was a secularised god. The ether took over where Nature left off. After the death of nature killed by natural history and industrial capitalism (see Giblett, 2004, chapter 1), along came the ether as the new god, the new god of the new superman with his body electric. The ether was the medium for both the spiritualistic and the naturalistic. The ether, Pocock (1988, p. 73) argues, 'was a common link between the natural and supernatural Orders' and 'the instrument of uniting the material and spirit worlds' (Jolly quoted in Pocock, 1988, p. 83).

The ether was not air or gas, though, like them, it was invisible. The ether was not the product of transformation of solid into gas; it was not the sublime by which modernity mastered nature (see Giblett, 1996, chapter 2), but the medium that pervades all matter, and all spirit. By mastering this medium, 'man' mastered matter and spirit in a happy and triumphal confluence of interest. What Early (1996, p. 326) calls 'the mysteries of communication on the unseen "ether"' was a means of unlocking the mysteries of matter and energy. It has also paved the way for communicating information. It also conquered space and time hence the droll portmanteau 'ethernity' espoused by decadent Alfred Jarry's (1911/1996, p. 104) *Doctor Faustroll*. Ethernity, and later electricity, Rosalind Williams (1990, pp. 100 and 103) argues, 'pemit[ted] domination over nature without the mediation of human labor' and 'realized the ancient ideal of an artificial infinity, a hushed, awe-inspiring refuge'. As external nature and the artificial infinite might arouse the sublime as they did for Burke (see Williams, 1990, p. 90), so electricity, in a word, was sublimity.

Electricity combines three aspects of sublimity: it is an ethereal phenomenon of nature; it is a means to master nature; and it is an artificial infinity. Sublime electricity was also shadowed, and made possible, by its necessary 'other', what Hughes (1983, p. 1) calls 'messy vitality'. Electricity embraced the full range of the great chain of being from primeval and lowly slime (see Giblett, 1996, chapter 2) to modern and highest sublime. The ether was a means of unlocking the mysteries of the very building blocks of nature. After quoting Lodge on the ether (see above) Pocock (1988, p. 70) goes on to argue that 'the ether was regarded as the basic material from which all matter and all energy was formed'. By 'discovering' and exploiting the ether, 'man' had become God because he could manipulate the basic matter and energy from which all other matter and energy was created. By mastering the key to spirit and matter, 'man' 'unlocked the secrets of nature'. The wireless, Czitrom (1982, p. 65) argues, 'had somehow put men on the threshold of the innermost secrets of nature'.

Yet the ether was only a hypothesis until 1887, the year Heinrich Hertz 'discovered' 'the invisible electromagnetic radiation which is the basis of radio telegraphy' (Pocock, 1988, p. 1; see also Aitken, 1976, pp. 48–79; Czitrom, 1982, p. 62). In the following year George Fitzgerald considered that

> Hertz's experiment proves the ethereal theory of electro-magnetism ... Fire, water, earth and air have long been his slaves, but it is only within the last few years that man [*sic*] has won the battle lost by the giants of old, has snatched the thunderbolt from Jove himself and enslaved the all-pervading ether.
>
> (quoted in Pocock, 1988, p. 70; see also p. 1)

By controlling the ether, 'man' is ostensibly able to control the other elements. After 'man' conquered the four other elements, the ether was the only one left. Yet electromagnetism was the heavenly power of the gods, the power of electricity represented by the thunderbolt and the lightning strike. By achieving victory in the heavenly realm where the giants had failed, 'man' made himself into a gigantic, gargantuan god.

Other writers put the Promethean snatching of fire from the gods much earlier. For Villiers de l'Isle-Adam (1963, p. 42), in the 1880s Benjamin Franklin, a century before, with his famous kite had 'snatched lightning out of the sky'. Elsewhere, de l'Isle-Adam (1981, p. 24) has his fictionalised Edison claim that 'since Franklin thunder has ceased to be divine'. But it did not cease to be sublime. Even though for de l'Isle-Adam's (1981, p. 24) Edison 'God is the most sublime conception possible', the death of God did not entail the demise of the sublime – quite the contrary. The sublime was secularised and the gods became mere metaphors for natural forces and ceased to be those forces themselves.

The sublime was thereby aestheticised. The sublime is a modern, secular theology, but it has its roots in ancient and medieval culture and society (see Giblett, 1996). Sublimation superseded magic; aesthetics superseded religious ritual. Writing about the waning of the Middle Ages, Huizinga (1924, p. 48) argues at that time

> all emotions required a rigid system of conventional forms, for without them passion and ferocity would have made havoc of life. By this sublimating faculty each event becomes a spectacle for others... The ceremonies accompanying birth, marriage, and death fully assumed this character of spectacles. Aesthetic values have here taken the place of their old religious (pagan for the most part) or magic signification.

Sublimation makes an event into a spectacle for spectators and transforms ritual into aesthetics. The ceremonies accompanying film and television festivals, premieres and prize-givings became spectacles. The ordinary, everyday events of stars' lives became spectacles for fans.

Prometheus bound for glory

By snatching the fire of electricity from the heavens, radio 'man' was the new unbound, modern Prometheus. Prometheus was 'a culture-hero' in Freud's terms who stole fire from the gods in the heavens and brought it down to earth, whereas the modern Prometheus is a culture hero who creates fire on earth. Nikola Tesla, an early pioneer of radio technology, became, as Hanson (1982, p. 4) puts it, 'Prometheus in reverse' by creating lightning. Not content merely to seize the power of lightning, Tesla wanted to create lightning like the Christian God, and even be Zeus, the Greek god of lightning. In a famous experiment conducted in 1899 in Colorado Springs he built a mast and a copper ball that looked like a lightning rod. Attached to the ball was a coil that generated millions of volts. When Tesla's assistant threw the switch, 'jagged streamers of lightning more than a hundred feet long and thick as a man's arm burst from the copper ball like God's own sneezes'. Although the overload blew out the main generator at the Colorado Springs Electric Company, Tesla had not only reversed Prometheus' act of stealing fire from the gods, but had become Zeus, the god of lightning.

Yet as Hanson (1982, p. 6) points out, Tesla was

> not the first to see this manifestation of natural forces as a symbol for intelligence and power. 'It is the thunderbolt,' wrote Heraclitus, a Greek philosopher of the sixth century B.C.[E.], 'which steers the course of all things.'

As cybernetics is the study of steering mechanisms 'from the Greek *kybernetes*, for steersman' (Wiener, 1954/1989, p. 15; Hanson, 1982, p. 51), electricity is cybernetic. By controlling electricity, 'man,' the 'God' of secular humanism, could control everything – or thought he could.

Yet radio did not merely bring the fire of the ether down to earth, or even create it; it brought the power of the ether into the home. Radio was what Czitrom (1982, p. 62) calls 'the ethereal hearth'. The old hearth of the fireplace as the centre of the private sphere where fire was used for cooking and heating presided over by Hestia/Vesta became the new hearth of the radio as the new centre of the private sphere ruled over by Prometheus where the new fire of the ether was used for informing and entertaining.

Radio was the first broadcast medium whose genesis dates from 1920 (Lax, 1997, pp. 24–25; Tucker, 1978, pp. 1253–1257). It played a vital role in the formation and maintenance of national identity. In 1922 there were 'an estimated million listeners and nearly 600 broadcasting stations in the United States' (Tucker, 1978, p. 1254). Broadcast radio in the United States, in Hanson's (1982, p. 35) words, 'brought the nation together as a single, vast audience; one that could be addressed intimately, yet simultaneously'. Although the audience was probably never as homogeneous or unitary as Hanson implies, radio may have been nation-forming without being nationalistic, or even tribe-forming.

McLuhan (1964, p. 298) argued that 'telegraph and radio neutralized nationalism but evoked archaic tribal ghosts of the most vigorous brand' pointing to Hitler's use of radio in Germany, where it operated as what McLuhan calls a modern-day 'tribal drum'.

Radio played a crucial role in the development of propaganda. 'Wireless telegraphy', Fuller (1961/1992, p. 138) argues,

> went far, if not to create psychological warfare, [at least] to give world-wide power to propaganda; to dement entire nations by transforming the spoken and written word into a weapon of war possessed of the velocity of light and in radius global. Further, it led to the development of the science of electronics.

Although Fuller recognized the importance of radio for propaganda, de Landa (1991, pp. 72 and 74) argues that he did not recognise the pivotal role of the radio in modern warfare, especially in the Second World War.

The Second World War has been labelled the 'Radio War'. Yet Moyal (1984, p. 153) states that 'this is inadequate. Radio and telephony... and telegraph[y] made it essentially the 'Telecommunications War'. Communication technologies have played a role in the lead-up to modern wars, and may even have precipitated them. According to Kern (1983, p. 275), 'observers during and after the First World War agreed that the telegraph and the telephone... were used more to bring on the war than to keep the peace'. When war broke out the latter was used to such an extent that the First World War 'has sometimes been called the "Telephone War" ' (Moyal, 1984, p. 153). It has also been called 'the Wireless War' when, as Hanson (1982, p. 32) puts it, 'this new form of electronic communication over the airwaves found immediate practical application as a weapon of war'.

The imperial powers, principally the Admiralty and War Office, were not only the first to make use of the radio for transmitting military and other information in the 1890s but also invested in its development (see Aitken, 1976, pp. 162, 164 and 167) and were its first customers (see Aitken, 1976, p. 232). The British Navy was unlike their transatlantic counterparts who were rather slow on the uptake because radio was both a command and control technology and a new communications technology that required a profound institutional realignment (see Douglas, 1985, pp. 117–173, especially p. 170).

Rather than the First World War, perhaps the Boer War was the first Wireless War or, at least, the first war for which radio equipment was purchased. Aitken (1976, p. 232) argues, 'the first actual sales of equipment by the Marconi Company were to the British War Office in 1898 for use in the Boer War', though I have been unable to find any evidence of its use. The Boer War was principally a telegraph war as we saw in a previous chapter. With radio, Smythe (1981, p. 304) argues, 'from the beginning the initiative was taken by the military (to communicate with naval vessels and posts on land) in the service of the imperial interests of the great powers'. Amongst the ways in which 'wireless went to war' (Hanson, 1982, p. 33) was by decoding enemy secrets. Wireless, Hanson (1982, p. 33) goes on to point out, 'was also the basis for a new branch of study – "traffic analysis," or cryptography, the practice of reaching into the airwaves to capture and decode wireless orders issued by the enemy'.

Radio was not just a new communication technology, but the key component in assembling Hilter's 'distributed network' of command and control in the Second World War that coordinated planes and tanks in *Blitzkreig*, literally 'lightning war'. De Landa (1991, p. 75; see also pp. 150 and 158; see also Cornwell, 2003, pp. 244–247 and 286 and van Creveld, 1991, pp. 190 and 267) goes on to argue that

> most German tanks and planes, in contrast with their Allied counter-parts, came equipped with two-way radio-capabilities. That is, they were conceived from the start as part of a network of arms, joined together by a wireless nervous system. In a sense *Blitzkreig* was not the name of a new tactical doctrine but of a new strategy of conquest, which consisted of terrorizing a potential target through air raids and propaganda and then breaking up its will to resist through a series of armoured shock attacks.

Radio was not simply a means of psychological warfare but an instrument of terror that was used both psychologically against civilians and militarily against enemy forces. *Blitzkreig*, for Virilio (2003a, p. 43), 'is no more than a kind of exhibitionism that imposes its own terroristic voyeurism: that of *death, live*'.

Electricity and the electromagnetosphere were the means to do so. The radio and the telephone, Hanson (1982, p. 35) argues, 'broke down geographic barriers, cut across time zones and loosened the isolation of small towns and communities' from the big cities and the seats of power and commerce. The electric spark (Czitrom, 1982, p. 9) was the flame that ignited 'the ethereal hearth' and set it ablaze with the sound of music and voices. The hills, and valleys, were alive with the sound of music, and voices. Electricity for Villiers de l'Isle-Adam's (1981, p. 69)

Edison was 'this astonishing vital agent' and for their contemporaries 'the Promethean fluid of vivification' (Michelsen, 1984, p. 8). Electricity was the martial power of the gods brought down from Mt Olympus, the home of the gods, to men on earth. By wresting the fire of electricity from the gods, 'man' procured their military might.

In Ovid's version of the Prometheus myth the eponymous hero steals the 'heavenly fire' of 'th'ethereal energy' from the 'God of Nature' and uses it to create a new man out of a mixture of earth and water (quoted in Hindle, 1992, p. xxiv; see also Bullfinch, 1993, p. 14 and Rose, 1928, pp. 53–56). Mary Shelley's *Frankenstein* is a cautionary tale of the dangers attached to doing so. It is usually read these days as a warning against biotechnology and genetic engineering. Yet it can also be read as a cautionary prophetic tale about the dangers of the communication technologies of cinema, radio and television that created a new man with a body electric. Rather than mixing the elements of earth and water, radio used the fifth element of the ether and its power of electricity to 'infuse a spark of life' (Shelley, 1818/1992, pp. 9 and 56) into modern man. Instead of 'infusing life into an inanimate body' like Frankenstein (Shelley, 1818/1992, p. 56; see also p. 51), radio, like cinema, infused electricity into an animate body. The unseen voice in the ether and the disembodied star on the screen are bodies electric.

In the dream that gave rise to the novel Shelley (1818/1992, p. 9) saw 'the hideous phantasm of a man stretched out, and then, on the working of some powerful engine, show signs of life, and stir with an uneasy, half-vital motion'. Jump forward a hundred years and the recording engine of the camera 'kills' the star on film only to have the projecting engine of the cinema bring it back to a half-vital life on the screen. Both engines are Bachelor Machines for Bachelor Births as is the engine of Shelley's Frankenstein (see Carrouges, 1975, p. 194; cf. Carrouges, 1954, p. 245). Like the star of the cinema, Frankenstein's creation is the dead living. By stealing the fire of electricity from the gods and creating new life, man became god.

By doing so, 'man' could exercise power over long distances. The power of communication is essentially the power to communicate over long distances. Communication is power. In Kipling's (1904, p. 219) story 'Wireless' the proponent of the new technology refers elliptically and haltingly to 'the magic [of]...the Hertzian waves...the Powers – at work – through space – a long distance away'. Power is constituted by the ability to act from afar. Communication technologies exercise power from afar; by acting from afar they eliminate distance and by eliminating distance kill.

Power in the sense of electricity is also characterised by the ability to act from afar. Maxwell's major contribution to the study of electro-magentism was to theorise electricity both as 'action at a distance' and as 'lines of force' (Smith-Rose, 1948, pp. 25–27). Radio's ability to act from afar by communicating messages along vectors over long distances employed the formal features of electricity and electromagnetism. Radio merely exploited 'natural' or physical resources.

Radio not only 'eliminated' distance like telegraphy and photography but also raised the possibility of creating and annihilating matter. Tesla (quoted in Pocock, 1988, p. 73) claimed that 'to create and to anni-hilate material substance, cause it to aggregate in forms according to his desire, would be the supreme manifestation of the power of Man's mind, his most complete triumph over the physical world'. As the ether pervaded the physical world, 'mastery of the ether' (Pocock, 1988, p. 73) was the means to do so and to extend his mastery over nature to the matter of which it was composed. Eighteenth-century landlords made the land, as Williams put it, 'move to an arranged design' into the 'pleasing prospects' of pastoral landscapes (see Giblett, 2004, chapters 1 and 3), whereas nineteenth-century physicists made matter move to their desires, made matter into pleasing forms according to their desires, made feminised matter into feminised forms that satisfied their desire for mastery. They exercised power of the mind over matter by giving birth to brain-children, and so achieved what T. Coraghessan Boyle calls 'the heady exhilaration of the victory over nature itself'.

Believers in the power of the ether saw it as magical. With radio, Miller (1997, p. 49) argues, 'the metaphysical and the material were linked: radio was the carrier of the magical-in-the-modern'. Wireless, for Beer (1996, p. 158), is 'domesticated magic'. Wireless breaks in and tames the wild ether making it into a pet or a slave. Wireless transforms the wild electromagnetosphere into the tamed private sphere. The boundaries of the domestic sphere are extended out into the electromagnetosphere, just as later the private sphere of the car (as we saw in the previous chapter) was mobilised in the public sphere of the street and the road against the global commons of the -ospheres.

The idea of 'the magical character of the Hertzian waves' (Early, 1996, p. 328; see also Kipling, 1904, p. 219) was commonplace in the late nineteenth and early twentieth centuries. Beer (1996, p. 154) argues that for Virginia Woolf writing in 1928 'technology has remade the world as magic'. In *Orlando* the eponymous central character thinks 'the very fabric of life now...is magic' (cited in Beer, 1996, p. 154). He/she 'listens to voices in America' which is not so much or any longer a real place but a virtual space/time of disembodied voices acting at a distance. As Johnson (1988, p. 11) sums it up, 'the magic, the marvel, the romance, and most frequently, the wonder of wireless were the terms in which commercial beginnings of this culture industry were hailed'. The modern is ostensibly anti-magical, or certainly against pre-modern magic; indeed, it defined itself against the magic of fairies and 'witches', but the modern is resolutely in favour of its own magic of the technological sublime.

The high priest, if not the god, of this secular religion was Thomas Alva Edison described by his contemporaries as 'the Magician of the Age', 'the wizard of Menlo Park' (Essig, 2003, p. 23), 'the man who took Echo prisoner' (de l'Isle-Adam, 1981, pp. xiii and 3) and 'the Faust of industrial capitalism' (Michelsen, 1984, p. 5) who sold his soul to the devil for power over nature and paved the way for radio, 'a subliminal echo chamber of magical powers', as McLuhan (1964, p. 302) calls it. Edison, for Metzger (1996, pp. 50 and 51),

> stands as the one man who might be said to have invented the twen-tieth century. Motion pictures, recorded sound, the telephone, the distribution of electric power and its corollaries (commercial lighting, household appliances, home entertainment), radio and public trans-portation were all shaped by Thomas A. Edison's work.

Like the car (as we saw in the previous chapter), radio is a magical object consumed by users who have little or no knowledge of its workings. Benjamin argued that 'the most precise technology can give its products a magical value' (1979, p. 243; 1999b, p. 510). This applies not only to commodities being fetishised in general by capitalist production and consumption as Marx argued in the first volume of *Capital*, but also to communication technologies in particular being used as instruments of power over time and space. What else is that if not magic? The essence of magic for Kafka (quoted in Virilio, 1989b, p. 41) was that it 'does not create, but summons'. The power of communication at a distance is to summon from afar, to make the other come to one instead of one going to the other. This is the essence of magic, and power.

One of the first writers about magnetism in the seventeenth century was Joan Baptista van Helmont (1650, p. 85), who saw it as a magical power with the ability to act at a distance. For him,

> in man there sits enthroned a *noble energy*, whereby he is endowed with a capacity to act *extra se* without and beyond the narrow

territories of himself, only *per Nutum*, by his single beck, and by the natural magic of his *Phansy* [from which magnetism proceeds] a subtle and invisible virtue … operate[s] upon an object removed at very large distance.

Electromagnetism was only to extend and heighten what van Helmont (1650, p. 88) later calls 'a power of acting … or moving any object at remote distance'.

Like telegraphy before it and as its successor, radio is a sublime communication technology. Radio transmissions on 'the transcendent waves on the aether' (Early, 1996, p. 328) communicated over insurmountable terrain and through bad weather. Radio surmounted the insurmountable and so was sublime as the surmountainous is one of its key features (see Giblett, 1996, chapter 2). Kipling's (1904, p. 213) story 'Wireless' begins with one of its characters proclaiming that 'nothing seems to make any difference … – storms, hills, or anything'. Radio was insensitive to distance, terrain and weather, to the horizontal and vertical dimensions of terrestrial space and the atmospheric conditions in it. Communication satellites were only to extend and heighten this distance-insensitivity into orbital extra-terrestrial space.

In reflecting back on the history of radio Smith-Rose (1948, pp. 31–32; see also Aitken, 1976, pp. 179–297) acknowledged its role in rocket and missile guidance, and speculated on its future in extraterrestrial transportation:

Marconi demonstrate[d] in 1896 the practical possibilities of wireless or radio signalling by means of electromagnetic waves freely transmitted through space or over the surface of the earth. Thus was born half a century ago one of the greatest amenities (albeit it is not without some disadvantages) of our modern civilisation, the ability to communicate with our fellow beings in all parts of the world whether on land, at sea or in the air. The resulting development and its widespread applications have given us all forms of radio communication, telephonic as well as telegraphic, with broadcasting and television, and material aids to aerial and marine navigation; also the ability, for good or ill, to control rapidly moving projectiles which can soar through the atmosphere to heights far beyond those ever attained by direct human exploration. The ultimate possibilities of this technique have yet to be envisaged.

These possibilities were to be realised with the launch of orbital extra-terrestrial satellites that initially emitted radio signals and then later received and transmitted them.

Radio began with, and in, the ether and reaches its height and breadth of extension with what Virilio (1989a, p. 112; 2000a, p. 145; 2000b, p. 20) calls 'the electronic ether of telecommunications' or, more recently 'the cybernetic ether' (Virilio, 2005b, p. 142). If the railway obliterated space, and telegraphy time, telecommunications combined both in what Virilio (1986a, p. 18; 1991b, p. 13) describes as 'an electronic ether, devoid of spatial dimension, but inscribed in the singular temporality of an instantaneous diffusion'. Radio was not simply the exploiter of the electromagnetic ether as a passive resource, but also the creator of its own telecommunications ether as an active agent.

The electronic ether of telecommunications creates what Virilio (1986a, p. 18; 1991b, p. 13) calls 'a new 'technological space–time' in which 'people can't be separated by physical obstacles or by temporal distances'. Live coverage of the 2003 Iraq War reticulated on monitors in the library and in the staff room of the university where I work and watched by a gaggle of students and colleagues attests to this electronic ethereal telepresence. Mapping this new telecommunications ether with the old communication lines of telegraphy and radio and the transport routes of canals, railways, highways and airways would produce what Virilio (1990, pp. 91–92) calls 'a tangle of networks' blackening the map. This blackened map of networks for him is 'the administration of a territory set up for the conductibility of war'. Territory is made open to visibility, to communication and to penetration, in brief, to war.

This continuum was the medium for what Takeguchi and Wooley (1992, p. 156) call the 'war in the ether' of the 1991 Persian Gulf War. This war has been dubbed 'the first information war' because of the use of computers and satellites, but the electromagnetosphere made communication between them possible. The electromagnetosphere became a front in warfare with the First World War, continued in the Second World War and culminated in the Gulf War. An information war, for Takeguchi and Wooley (1992, p. 155), is 'fought as much in the *airwaves* as it is on land, sea, air and space'. Commenting on the 1991 Gulf War, Campen (1992a, p. xi) insists that 'it is simply not possible to wage information warfare without assured dominance in the electromagnetic spectrum'. And without dominance in orbital extraterrestrial space it is simply not possible to wage and win information warfare either.

8
Disciplinary and Flânerie: The Panopticon and Panorama of Television

Television arose out of radio and film: television is basically 'sending pictures by radio' (Hanson, 1982, pp. 35–6), or simply radio with pictures. Yet the development of television was far from simple, and it was slow, though it had been imagined for a long time. In his brief and useful history of television Tucker (1978, p. 1257; see also Lax, 1997, pp. 37–39) argues that 'the desire to transmit pictures rapidly to a distant place by some electric-telegraphic method was evinced almost from the beginning of electric telegraphy', indeed from as early as 1850. The reason for the long delay between the invention of radio and that of television was because the latter attracted little investment. Raymond Williams (1974, p. 11) relates that 'a system of television was foreseen, and its means were being actively sought but also…, by comparison with electrical generation and electrical telegraphy and telephony, there was very little social investment to bring the scattered work together'.

The reason why there was little investment, social, economic or scientific, in the development of television is because there was no perceived immediate business or military applications or uses for television. As an extension of broadcast radio, television was seen as a technology of leisure, not of business or war. Television is an interesting kind of counter case to those technologies, such as radio and telegraphy, which were taken up and developed for use in war. The development of radio transmitters and receivers and, as Smythe (1981, p. 305) puts it, 'the civilian uses of the radio [electromagnetic] spectrum were a spin-off from war-financed research and development for military purposes'. Unlike radio, there was no initial military or business investment in the development of television. By virtue of *not* being selected for business or military investment, television demonstrates the power of the industro-military complex as a driver of technological development. If television had been seen to have immediate business or military uses, it would have attracted business or military investment and been developed much earlier.

Moreover, if aerial photography (and cinematography) had not already been in use for military purposes, television would have been developed much more quickly for aerial reconnaissance. And if radio had not been developed for point-to-point communication and then became a broadcast medium, television would have been developed much more quickly. The fact that radio had become a broadcast mass medium and that television was radio with pictures meant that television was seen initially as a broadcast medium. Military support could both advance the development of one communication technology and impinge that of another. For Winston (1998, pp. 96 and 97), 'the same imperial arms race, which necessitated the development of the marine wireless, suppressed work on television because no military application could be envisaged'.

Whether or not the military or business supported or did not support the development of a communication technology depended on social and cultural factors. For Raymond Williams (1974, pp. 11 and 12),

> the critical difference between the various spheres of applied technology can be stated in terms of a social dimension: the new systems of production and of business or transport communication were already organised, at an economic level; the new systems of social communication were not.

In other words, communication technologies were initially used for existing purposes, such as business communication. Communication was a means to an end; communication initially was not a business in its own right and the communication technologies were for business, not for pleasure. They were used initially for one-to-one business communication and not for one-to-many broadcast communication to a mass audience.

Yet the new systems of social communication were not developed evenly and uniformly at the same speed. Telegraphy, radio and satellites were developed much more quickly than television because they were seen to have advantageous military, business, imperialist and administrative uses. Institutional uses came first before individual use in the case of one-to-one communication by telegraphy and satellites. Institutional uses also came before commercial or national broadcast use in the case of one-to-many communication via radio and satellites.

By *not* attracting investment, television puts paid to 'technological determinism'. This is the idea that technology determines its own development, that technology develops determined by technological imperatives and not by social and cultural forces, especially institutional ones such as business and the military (see Lax, 1997, pp. 108–111 and Williams, 1985, p. 129; 1989, p. 120). This is not to deny, however, that technology has effects – quite the contrary. Rather it is to suggest that the social and cultural factors and forces that produced the technology in the first place produce these effects. Technology encapsulates and encodes social and cultural forces that are then transported and later decoded by the use of the technology. Technology is a relay between one set of social and cultural factors (which give rise to it) and another set (which reproduce them). This culturally determinist view of technology not only sees culture determining technology but then also sees technology determining culture in both a linear process and feedback. The very forces and factors that determine the technology are reproduced in and by the technology, which then contribute to the determination of culture. Technology is not neutral, but both a cultural product and cultural producer and reproducer.

Television is a case in point as Fisher and Fisher (1996, p. xvi) relate that it 'did not arise like Venus, springing newborn and whole from an oyster shell' as in Botticelli's painting. Nor did it arise like Athena, springing fully formed and armed from the head of a single, Zeus-like inventor however much it was a Bachelor Birth from a Bachelor Machine. Rather its appearance was the product of a century-long gestation and parturition at which many men laboured. They tried 'to pull electric visions out of the air' and eventually 'sent moving pictures through the air'. Television was elusive, the Fishers go on to relate, as it 'sidled up to us from a corner, then receded into the mists that obscure the future' until 'finally it was dragged kicking and screaming out of the mists, out of the theoretical uncertainties and technical difficulties that had masked and disguised it, and was made to work'. Engineering communication theory was the very proper and discreet girl that not only accepted your telegram but also ended up delivering television – however reluctantly. Television itself was the youthful goddess of information performing her provocative belly dance right on the screen in front of you in your own home. She had a difficult birth, though, as some sort of she-devil who had to be dragged kicking and screaming out of the mists of the heady heights of scientific theory.

By arising out of radio, television began life as a broadcast communication technology. On the other hand, by addressing the sense of sight it had more to do with cinema than radio, yet like radio it was received within the private domestic sphere of the home. Television brought cinema home; it brought the power of seeing that cinema enabled into the lounge room. Like cinema and the car, it was associated with a powerful shift in the logistics of perception. For Virilio (1988, p. 118), 'the new windshield is no longer a car, it is a TV screen'. Like cinema and the car, television was also associated with a powerful shift in the logistics of perception, and locomotion. Following Raymond Williams, Inglis (1995, p. 53) argues that 'the geographical mobility of the car and the cognitive mobility of the television have together driven people back into their smallest social compass, making them spectators of everywhere, and active nowhere'.

Television is what Virilio calls 'a vehicle for seated passengers' that transports us, or at least our gaze, to distant places whilst confining us in front of the TV set. Television empowers the eye but disempowers the rest of the body reducing it to an immobile passenger seeing everything but going nowhere. In this chapter, I argue that television is both mobilising and immobilising at the same time: television enables the gaze of the viewer to travel virtually to many places whilst the body of the viewer is positioned in frozen, immobile space in one place in the leisure-discipline of industrial capitalism. Television, in other words and in short, is both panorama and panopticon. Using the work of Benjamin, Foucault and others, this chapter relates these two contemporaneous 'machines of visibility' to the paradoxical disciplinary *flânerie* of television. Television, in Virilio's terms, is a third window through which light passes, but not bodies, whereas what Richard Jefferies called 'the fourth window' open to the stars enables both light and bodies to pass through.

Panopticon

Television produced new ways of seeing and inherited ways of seeing and modes of representation that were endemic to modernity. These entailed a non-reciprocal relationship between seer and seen: the seen could not see the seer. The architectural (largely conceptually as it was not built), and modern archetypal, configuration of this asymmetry was Bentham's (1787/1962, pp. 35–170) panopticon made famous by Foucault (1977, pp. 200–209; see also Thompson, 1995, pp .132–134). Although this model prison is a useful device for explaining some of the features of television (as I will try to show shortly), its applicability to modern media has been the subject of some contention. Thompson (1995, p. 132) argues that

> Foucault did not discuss directly the nature of the media and their impact on modern societies... If Foucault had considered the role of communication media more carefully, he might have seen that they establish a relation between power and visibility which is quite different from that implicit in the model of the panopticon.

Whilst Foucault did not consider the role of communication media extensively, he did consider them albeit in passing, and not just in relation to the panopticon. He argued (1977, pp. 170 and 171) that 'the techniques that make it possible to see induce effects of power'. He goes on to discuss how 'the major technology of the telescope, the lens and the light beam' converged in the communication technology of cinema.

Alluding to the *camera obscura*, Foucault (1977, p. 207) argues that the importance of the panopticon lies in the fact that 'the seeing machine was once a sort of dark room into which individuals spied; it has become a transparent building in which the exercise of power may be supervised by society as a whole', and not just in and by the communication media. For Foucault (1977, p. 216), 'the modern age poses the... problem: "To procure for a small number, or even for a single individual, the instantaneous view of a great multitude"'. Conversely, the hypermodern age of the media, for Thompson (1995, p. 134), poses the problem of procuring for the great multitude the instantaneous view of a small number of stars:

> Whereas the panopticon renders many people visible to a few and enables power to be exercised over the many by subjecting them to a state of permanent visibility [or more precisely, to the threat or possibility of it to the point that they internalise it (see Bentham 1787/1962, pp. 40 and 44; Foucault, 1977, pp. 187 and 201)], the development of communication media provides a means by which many people can gather information about a few and, at the same time a few can appear before many; thanks to the media, it is primarily those who exercise power, rather than those over whom power is exercised, who are subjected to a certain kind of visibility.

This is a remarkably sanguine and congratulatory view of the 'public service' role of the media. If the media serve a useful role (as they do) in making the few visible to the many, why isn't the power of the media to

and other residents similarly positioned in their frozen, immobile and segmented private spaces right next door, across the road or down the street. Foucault (1977, p. 197) remarks that 'this enclosed, segmented space...in which the individuals are inserted in a fixed place...constitutes a compact model of the disciplinary mechanism'.

Rather than any strict similarity or correlation between the spatial practices of the panopticon and of watching television, it is the deployment of panopticism as a disciplinary mechanism in both that is pertinent. Rather than the panopticon *per se* being an apposite model for the power of communication media, panopticism is the principle of their *modus operandi*. Rather than narrowly focussing on the panopticon, Foucault is really only concerned with it as an instance of panopticism. After all, his account of the panopticon only begins five pages into his discussion of panopticism.

Like the railway, the car and the cinema, panopticism immobilises the body and mobilises the gaze. The panopticism of cinema and television (and later computer monitors and multimedia) constitutes the triumph of the eye over the hand in representation and in discipline. In the late eighteenth century Foucault (1973, p. xiii) argues that the power of the gaze becomes predominant because

the eye becomes the depositary and source of clarity; it has the power to bring a truth to light that it receives only to the extent that it has brought it to light; as it opens, the eye first opens the truth: a flexion that marks the transition from the world of classical clarity – from the 'enlightenment' – to the nineteenth century.

This transition is made via the panopticon of the late eighteenth century to the photography of the early nineteenth century. The eye brings the truth to light. Light itself is insufficient to do so.

The power of the eye to see the truth was linked to the new power of discourse to say the truth. At the beginning of the nineteenth century Foucault (1973, p. xii) argues that 'a new alliance was forged between words and things, enabling one *to see* and *to say*'. Foucault (1970, pp. 135 and 157–162) associates the beginnings of this alliance with the development of the 'discourse of nature' involving observation (seeing linking spontaneously with saying). Just as, in Foucault's (1970, p. 160) terms, 'nature is posited only through the grid of denominations without [which]...it would remain mute and invisible' so the prisoner in the panopticon is posited through a grid of spatial denomination and is made visible and to speak, or at least to signify by the play of light on the surface of his body, by the forms and postures he assumes.

Just as nature became visible and was made to speak through and by the denominative grid of taxonomy, so the (human) body became visible and was made to speak or signify through and by the denominative grids of panopticism. Power, for Foucault (1977, p. 202),

> has its principle not so much in a person as in a certain distribution of bodies, surfaces, lights, gazes; in an arrangement whose internal mechanisms produce the relation in which individuals are caught up.

The distribution of bodies, surfaces, lights, gazes was to achieve its modern apotheosis in the new advertising pictorialism of the late nineteenth century and in the birth of the new cinema star in the early twentieth century. The power of panopticism does not reside in the individuals who operate (in) it, but in the positions that they are made to take up in it and in the relations of visibility or invisibility and the lines, or vectors, of communication between them. Both come together in the gaze which, for de Certeau (1987, p. 15), 'is a vector – a line and an action in space'. The line of sight (and fire) aims to eliminate distance and exercise power. Panopticism is the vectorisation of the gaze.

The panopticon is just one instance of these relations of visibility or invisibility and of vectors of communication. Architecturally it is a spatial configuration with an arrangement of juxtaposed windows. Yet the inner window facing the guard's tower is more precisely what Virilio (1988, p. 191) calls the 'door-window' that allows both light and bodies to pass through. For Virilio this is the first window. He goes on to differentiate it from the second window, the window in the outside wall of prisoners' cells. This window is 'the window which gives us the day – only light – a window through which one doesn't enter' as a rule, and can't leave as in the panopticon. This window produces backlighting that, for Foucault (1980, p. 147; see also 1977, p. 200 and Thompson, 1995, p. 133), 'enables one to pick out from the central tower the little captive silhouettes in the ring of cells'. With the second window for Virilio 'what enters through it is only an abstraction – the sun, direct vision' just as sun enters through the lens of the camera and is recorded in the frame of the film to be reproduced in the photograph and the cinema.

The third window, for Virilio (1988, pp. 191–2), is the screen window such as the cinema screen, 'the cathode-ray window' of television (Virilio, 1986a, p. 21; 1991b, p. 17; see also 2001, p. 69), the car windscreen and Windows®. He argues that 'it doesn't function at all as a medium like radio or newspapers, but as an architectonic element: it is a portable window' just like the laptop and personal digital assistant. After the medium comes the screen. After the cell of the panopticon and its static, frozen, immobile space comes the static, frozen, immobile body of the screen jockey gazing fixedly at the screen monitor wherever his or her body happens to be at the time. Virilio (1993, p. 4) argues later that

> the development of territorial space by means of heavy *material* machinery (roads, railways and so on) is now giving way to an almost *immaterial* control of the environment (satellites, fibre-optic cables) that is connected to the terminal body of the men and women, inter- active beings who are at once emitters and receivers.

These beings are, in a word, 'cyborgs'. What Virilio (1993, p. 5) calls 'the *static audiovisual vehicle*' with its associated 'behavioural inertia of the receiver–sender' leads to 'the 'bodily suspension' of the 'plugged-in human being' and 'the immobile traveler' (Virilio, 2000e, p. 106). Television, Virilio (1993, p. 11) concludes, no longer requires 'human mobility, but merely a local motility'. From the panopticon to the television there is what Virilio (1993, p. 11) calls a '*growing inertia*' and the '*inertia of one's own body*' (Virilio, 2000b, pp. 46, 47, 57 and 60) 'installed in front of the cathode display cases' of the museum of accidents, the stuff-in-trade of the CNN coverage of the Gulf War of 1991 and the World Trade Centre bombing of September 11, 2001.

There is, however, a fourth window that Virilio (2007, p. 48) mentioned recently when he argues that 'contemplation of a screen not only replaces contemplation of script, ... but also contemplation of the stars' at night. The fourth window is the window to the stars above in the night sky (rather than the stars below of the screens). This window is perhaps more precisely what Richard Jefferies (2001, pp. 20 and 21), a late-Victorian nature writer writing in the 1880s, calls the 'window of the trees' through which he sees the stars by night and the sun by day when he is lying on his back on the ground (Jefferies, 2001, pp. 17–21). Unlike the third window of the cinema and television screen, and of the car windscreen and windows, unlike the panorama and the landscape painting (and painter) and unlike the sublime landscape or seascape (see Giblett, 2004), all of which are set up vertically in front of the seated or standing viewer, the fourth window to the stars and sun is placed horizontally above the supine viewer.

Moreover, unlike the second window of the panopticon through which light passes (but not bodies), the fourth window to the stars and sun is like the first door-window of the panopticon through which light and bodies pass. Yet unlike the imprisoning panopticon where the prisoner passes through the first door-window at the whim of the institution, Jefferies' body (and not just his mind) is liberated and sublimated through the fourth door-window as he is transported to the stars without leaving the earth (and his body). He relates how 'I walked amongst the stars. I had not got then to leave this world to enter space: I was already there' (p. 18). Jefferies is bodily and simultaneously on earth and in the heavens. He walks amongst the stars, not as a fan, but as a friend. The windows to stars were 'loopholes in the trees' through which he travels in space (p. 20). He sublimates both body and mind, not body into mind (see Giblett, 2004 and forthcoming).

Although Foucault did not address explicitly modern communication media or acknowledge the pertinence of his work on panopticism to them, others have since done so such as Friedberg (1993, p. 20), who suggests that

> like the central tower guard [in Bentham's panopticon], the film spectator [and television viewer] is totally invisible, absent not only from self-observation but from surveillance as well. But unlike the panoptic guard, the film spectator [and television viewer] is not in the position of the central tower, with full scopic range, but is rather a subject with a limited (and preordained) scope.

As such, they are just like the prisoner in the cell. Yet unlike the prisoner, the film spectator and television viewer is not placed under surveillance but views politicians and stars placed under surveillance by the disciplinary mechanism of the communication technology. The film spectator and television viewer are both in the tower (in a position to see and not be seen) and in the cell (in a segmented, frozen and immobile space). The limited and preordained scope of the film spectator's and television viewer's position is what Friedberg calls 'synoptic' rather than panoptic. Cinema and television are the synopticon, rather than strictly the panopticon.

The synopticon entails what Benjamin (following Baudelaire) calls *flânerie*, whereas, as we have seen, the panopticon is disciplinary. Benjamin (Adorno and Benjamin, 1999, p. 310) defines *flânerie* as 'a state of intoxication', whereas the disciplinary can be defined as a state of incarceration. Friedberg (1993, p. 16) goes on to suggest that the trope of *flânerie* delineates a mode of visual practice coincident with – but antithetical to – the panoptic gaze. Like the panopticon system, *flânerie* relied on the visual register – but with a converse instrumentalism, emphasising mobility and fluid subjectivity rather than restraint and interpellated reform.

Friedberg draws too sharp a distinction between the two: cinema and television give the eye power to wander virtually and imaginarily over scopic vista, but position the viewer in a seat inside an interior space. Similarly, the car not only gives the driver and passenger mobility to travel and view scenes and landscapes, but also positions them inside a metal box, a communication prison. The panopticon immobilises physically the body and the gaze of the inmate, and mobilises the gaze and body of the guard, albeit within a narrow round; the synopticon of cinema and television mobilises the gaze of the spectator and viewer, but immobilises their body.

The synopticon, for Friedberg (1993, p. 22), is the direct descendant and inheritor of the panorama, rather than the panopticon. The panorama, Friedberg (1993, p. 22) relates, was

> a 360-degree cylindrical painting, viewed by an observer in the center...The panorama did not physically *mobilize* the body, but provided virtual spatial and temporal mobility, bringing the country to the town dweller, transporting the past to the present.... The panorama offered a spectacle in which all sense of time and space were lost, produced by the combination of the observer in a darkened room (where there were no markers of place or time) and presentation of 'realistic' images of other places and times.
> (see also Schivelbusch, 1986, pp. 61–62; 1988, pp. 213–221; Kirby, 1997, pp. 7 and 42–48; and Oettermann, 1997)

The panorama, like the panopticon, was a cylindrical space with the observer placed in the centre in a darkened space. Yet unlike the guard in the panopticon whose body and gaze were mobilised to place the prisoner potentially under surveillance, the viewer of the panorama had their body immobilised and their gaze mobilised.

Bentham's panopticon and Robert Barker's panorama were contemporaneous. Barker received a patent for his invention in 1787, though he did not use the term 'panorama' when he applied for a patent earlier that year (Oettermann, 1997, pp. 5 and 6). This was the same year in which Bentham conceived his panopticon. Yet the connection between

Bentham's panopticon and Barker's panorama may be more than mere coincidental contemporaneity. In a note to his discussion of the panopticon Foucault (1977, p. 317, n.4) wonders whether

> Bentham [was] aware of the panoramas that Barker was constructing at the exactly the same period (the first seems to have dated from 1787) and in which the visitors, occupying the central place, saw unfolding around them a landscape, a city or a battle. The visitors occupied exactly the place of the sovereign gaze.

The panoramas positioned viewers to gaze upon and master the monumental.

The panorama, for Oettermann (1997, p. 7), was 'a kind of pattern for organizing visual experience' in which 'real landscapes were experienced as (artificial) panoramas, and the panoramic view of landscapes became the dominant mode of seeing'. Land that did not conform to the panoramic was found to be wanting just as it had been for the aesthetic, painterly and poetic modes of the beautiful, picturesque and sublime before it. The panoramic entailed an aestheticisation of the land and views of it into landscape, as did the beautiful, picturesque and sublime (see Giblett, 1996; 2004). The panorama, for Oettermann (1997, p. 7), was 'the pictorial expression or "symbolic form" of a specifically modern, bourgeois view of nature and the world'.

The panorama also involved an industrial capitalist technologisation of land as landscape. It did so in arguably the first mass medium, certainly the first mass visual medium. The panorama, Oettermann (1997, p. 45) argues later, is 'the art form of the Industrial Revolution'. He goes on to argue that

> For the first time art and artists found a patron in the masses. And the panorama, in turn, was the first art form to attempt to fulfill the visual needs and desires of anonymous city dwellers.

Cinema was to inherit, and supersede, this feature of the panorama.

The panoramic entailed an industrial capitalist asetheticisation and technologisation of not only the land but also vision. The pictorial panorama, for Oettermann (1997, p. 7), was

> in one respect an apparatus for teaching and glorifying the bourgeois view of the world; it served both as an instrument for liberating human vision and for limiting and 'imprisoning' it anew.

By mobilising the gaze (and not the rest of the body) of the viewer, the panorama is the direct descendant of the railway (Schivelbusch, 1986; pp. 52–69 Cubitt, 1998, p. 30; Williams, 1982, p. 74) and the direct antecedent of the cinema and television. For Oettermann (1997, p. 44) too 'the television of today is a direct descendant of the panorama'. He goes on to argue that

> The human eye was first schooled for the huge task of keeping the body immobilized in the panorama, while in its counterpart, the panopticon, the psyche was trained to submit to the tyranny of the clock.

And so to the time-discipline of industrial capitalism (Thompson, 1993), both the work-discipline and leisure-discipline, and to the modern urban partitioning of space and the sequestering of individuals.

In what Oettermann (1997, p. 45) calls 'the dialectic of the bourgeois mode of seeing', 'the sight of the panorama set people in motion [or at least their gaze], motion that followed the rhythm of factory machines' whilst immobilising their bodies like the prisoners in the cells of the panopticon.

The panopticon and the panorama mobilised the gaze whilst immobilising the rest of the body unlike the *flâneur*, whose gaze and body is mobilised like the guard's in the panopticon. Yet, despite this difference, both the panorama and the *flâneur* shared a common techno-pastoral aesthetic of the garden in the machine (see Giblett, 2004). Benjamin, for Friedberg (1993, p. 23), 'saw a direct relation between the panoramic observer and the *flâneur*: "the city-dweller... attempts to introduce the countryside into the city. In the panoramas the city dilates to become landscape, as it does in a subtler way for the *flâneur*." ' In the machines of cinema and television the city becomes landscape to be viewed in the same way that the train traveller, or tourist, viewed the country as an aesthetic, panoramic and pastoral object. Television also displays the commodities of consumer culture via advertising in what Crowley and Heyer (1991, p. 216) call its 'panoramic view' of the garden of earthly delights in the machine of sublime heights.

The *flâneur* is master of city and country because he is master of the passing parade of both, whether it be the actual life of the city streets through which he wends or the virtual life of the country which he sees in the panorama. Both mobilise the gaze. Baudelaire (cited in Benjamin, 1999a, p. 443) maintained that 'for the perfect *flâneur*, ... it is an immense joy ... to be away from home, yet to feel oneself everywhere at home; to see the world, to be at the center of the world, yet to remain hidden from the world... a kaleidoscope endowed with consciousness'. Like the guard in the panopticon, the *flâneur* is at the centre of the world, yet hidden from the world. He is at home everywhere, but is nowhere at home; his home is nowhere, just like the cinema-goer and television viewer; not nowhere in the sense of utopia, or eutopia, but atopia, no place.

The gaze, for de Certeau (1987, p. 20), 'abolishes every position that would guarantee the traveller an acceptable place, an autonomous and sheltering dwelling, an individual and objective "home"'. The gaze destroys home and produces a virtual traveller. As de Certeau (1983, p. 26) argues elsewhere, 'to see is to travel, but seeing is already travelling'. Television is such a perfect modern machine because it is not only a disciplinary mechanism that partitions space and positions viewers, but also a mechanism of *flânerie* in which the viewer, though fixed in an immobile space, has a virtual mobile gaze that enables him or her to, like the *flâneur*, 'botanize on the asphalt', wherever the camera takes his or her gaze and wherever the sound recorder takes his or her aurality (aural acuity). Television is a paradoxical modern machine, like cinema and the car, as it mobilises the gaze and immobilises the body at the same time in what Virilio (2005b, p. 99) calls 'the sedentary exhibitionism of television'.

Fan

The media *flâneur* is a fan; the fan is a *flâneur* of the media and media events. Thompson (1995, p. 222) traces how 'the term itself is an abbreviation of "fanatic" and was probably first used in the late nineteenth century to describe enthusiastic spectators of sport' and goes on to suggest that it has not entirely lost its connotations of 'religious fervour, frenzy and demonic possession'. If the star is the heavenly body of modernity sublimated into the cool ethereal and sidereal realms of the dead living, then the fan is its, or his or her, hot, hellish counterpart in the land of the modern melancholic living. The star is sublime to the fan's slime. Both are coupled in the loving and deadly embrace of Sofoulis' parenthetical portmanteau s(ub)lime (see Giblett, 1996, chapter 2). The fan is the hellish body of modernity basking and/or shivering in the black sun of the dead living star.

The star is sublimated whilst the fan is desublimated. After the star's body is sublimated into the image and then commodified in the sublimate of technological reproduction, the fan buys and consumes this commodity with his or her own body. The commodity begins to disintegrate as it is re-embodied and starts to fall back to earth, the earth from which it came, becomes again the raw material from which it was made, becomes bodily. The elements of the secular transubstantiation that saw the flesh and blood of stars transmuted into an unholy sacrament initially in the cinema cathedrals are desublimated back into the messy body of the fan plagued by capitalist desire, wracked with commodity pleasure and wallowing in the slough of despond of modern melancholia, the ruling humour of modernity (see Giblett, 1996). From this nadir the whole process gets going again to sublimate solid, base matter into airy, heavenly realms thus completing a full circuit in the 'psychogeocorpography of modernity' (see Giblett, 1996, chapter 2).

The fan is both *flâneur* and collector. For Benjamin (1999a, p. 207), 'the *flâneur* [is] optical, the collector tactile'. After photography disjoined the eye from the hand in pictorial reproduction so that the eye could wander freely as a *flâneur* (the *flâneur* is primarily a wandering eye), industrial production rejoined the eye to the hand so that the collector fondles the commodity fetishes of his or her collection. Benjamin (1999a, p. 206) argues that 'the collector's items as such were produced industrially'. The collector is a creature of industrial capitalism. The fan is a collector who collects 'media products', as Thompson (1995, p. 222) indicates.

Between the shiny heavenly star and the slimy hellish fan lies the earthly full body of the media magnate who mediates both and makes both possible. 'Some kind of full body,' Deleuze and Guattari (1977, p. 11) argue, 'that of the earth or the despot, a recording surface..., a fetishistic, perverted, bewitched world are characteristic of all types of society as a constant of social reproduction'. I would add the film, the disc and the page to the list of recording surfaces with a full body, and to the despot I would add the magnate and the mogul, the CEO and company chair. The media magnate writes the star and the fan on the full body of the media text. Orson Welles' *Citizen Kane* is the best fictive and filmic portrayal of the media magnate as a full, perverted body who reconstructs his Xanadu as a full body of the earth transforming a Florida desert into a private pleasure ground full of plundered living beings and collected dead things.

Out of this process the earthly magnate produces the heavenly body of the star and the hellish body of the fan. For Deleuze and Guattari (1977, pp. 11–12), 'the essential thing is the establishment of an enchanted recording or inscribing surface that arrogates to itself all the productive forces and all the organs of production, and that acts as a quasi cause by communicating the apparent movement (the fetish) to them'. The recording surfaces of the media text onto which both the star and the fan are inscribed are produced ultimately from the earth and are inscribed on the earth just as Kane rewrites a Florida desert into his Xanadu. The earth, for Deleuze and Guattari (1977, p. 141), is

> the element superior to production that conditions the common appropriation and utilization of the ground. It is the surface on which the whole process of production is inscribed, on which the forces and means of labour are recorded, and the agents and products are distributed.

Without the earth, no fan and no star, no hell and no heaven. The economic base rests on the earth (see Giblett, 2004). It furnishes the chemical elements for the process of sublimation, for the application of the heat of fire.

The fan is not powerless, but possesses and exercises power over himself or herself and over his or her body. Being a fan, in Foucault's terms, is a technology of the self which permits

> individuals to effect by their own means or with the help of others a certain number of operations on their own bodies and souls, thoughts, conduct, and way of being, so as to transform themselves in order to attain a certain state of happiness, purity, wisdom, perfection, or immortality (1988, p. 18; see also Thompson, 1995, p. 210).

In other words, sublimate themselves into the realm of heavenly bodies along with stars. But by doing so, or trying to do so, the fan constitutes himself or herself as the dead living like the star whose living body is deadened by the commodified image of it.

Technologies of the self are, for Foucault (1988, p. 18), one of four technologies that 'hardly ever function separately'. The other three are as follows:

> (1) technologies of production, which permit us to produce, transform, or manipulate things; (2) technologies of signs systems, which permit us to use signs, meanings, symbols, or signification; (3) technologies of power, which determine the conduct of individuals and submit them to certain ends or domination, an objectivising of the subject.
>
> (1988, p. 18)

The media are an instance of the four technologies operating together. Yet the various aspects of the media can be associated with one particular technology rather than another. The star is primarily a technology of the sign though he or she, or it, is also produced by technologies of production, reproduces himself or herself by technologies of the self and has power exercised over him or her by technologies of power, and exercises power over the fan. The star is a symbol, both a product of a technology that produces symbols and a producer of symbolic meanings that the fan reproduces.

The media magnate is primarily a technology of power, though he or she, or it, also produces by technologies of production, reproduces himself or herself by technologies of the self and makes meanings with technologies of the sign. He or she determines to a large extent the conduct of individuals employed within his or her media empire through senior management and editorial appointments and to a lesser extent the conduct of the consumers of his or her media products, the fans of his or her galaxy of stars, through the protocols of editing and scheduling and the demographics of audiences and markets.

The fan is primarily a technology of the self as I have suggested, but he or she is also produced by technologies of production, makes meanings with technologies of the sign and has power exercised over him or her and exercises power over others through technologies of power. The media apparatus is primarily a technology of production that produces programmes (television and computer), films, CDs and DVDs to be sold and bought, consumed and collected. Yet it also mediates power through the technologies of power, conveys meaning with the technologies of sign systems and makes it possible for the technologies of the self to operate within its domain. The media apparatus colonises and encloses not only terrestrial space via the full body of the magnate but also extraterrestrial space via satellites orbiting in the sublime company of heavenly bodies.

9
Orbiting in the Sublime Company of Heavenly Bodies: Satellites from Cold War to Gulf War

The anniversaries of the launch of the first satellite in 1957 pass by largely unnoticed, except as part of the celebrations of the conquest of orbital extraterrestrial space it began rather than for its contribution to communications. Although the world-shattering significance of the launch of the first Sputnik in 1957 may escape us, it did not escape the notice of contemporary reporters and philosophers at the time. Writing in the following year, Arendt (1958, p. 1) described it as an event 'second in importance to no other, not even to the splitting of the atom'. Arendt noted that the immediate contemporary reaction was not 'pride or awe' at this display of 'power and mastery' over nature, but what one American reporter called 'the first step toward escape from men's [sic] imprisonment to the earth' (quoted in Arendt, 1958, p. 1). Rather than acknowledging the power of 'men' over nature displayed yet again by the Sputnik, the earth was seen in a massive act of disavowal as having power over 'men' to imprison them. The mastery of 'men' over nature was justified by what they saw as the earth's mastery over them.

Escaping the prison of the earth

In this scenario, the Sputnik was seen as the means for 'men' to escape the smothering clutches of the mothering earth and to orbit with the Sputnik in what Arendt (1958, p. 1) called 'the sublime company' of other 'heavenly bodies'. The first step in escaping from the imprisonment of 'men' to the earth is sublime as it represents the desire to escape the solidities of the earth into the heights of the gaseous heavens. The chemical process of transformation of a solid into a gas is pressed into the metaphorical service of the philosophical sublime. The satellite made physically possible the first step in what philosophers from Socrates to

Kant had only dreamed about: escaping the 'prison of the earth' (see Giblett, 1996, chapter 2).

Satellites may be the first step in escaping the prison of the earth using the technological sublime, but the technological sublime itself is imprisoned in an earthly, masculine poetics and politics. The launch of the first Sputnik took place in what Arendt called the 'uncomfortable military and political circumstances' of the Cold War (see also Lax, 1997, pp. 17–18). Sputnik was about the size of a beach-ball, but despite its size de Landa (1991, p. 170) describes it as '184 pounds of orbiting paranoia'. It changed the nature, and theatre, of war. The Cold War, for McLuhan (1964, p. 339), was 'really a battle of information and of images fought by informational technology that goes far deeper and is more obsessional than the old hot wars of industrial hardware'. It was yet another instance of the way in which for him 'all wars have been fought by the latest technology available in any culture' – whether it be the use of the railway in the Civil War, as McLuhan argued, as we have seen previously (chapter 2), or the use of satellites in the Cold War.

Yet whereas the railway was a sublime technology that used heat to transform matter into energy, the satellite and information technology sublimated solid matter and hot energy into cold information. As the railway colonised and militarised terrestrial space, so the satellite colonised and militarised extraterrestrial space. What McLuhan (1964, p. 342) called 'this trend toward more and more power with less and less hardware that is characteristic of the electric age of information' was to intensify with the Vietnam and Gulf wars with the use of more and more software in surveillance and targeting technologies.

The Cold War took war to new heights and used new means of terror – on both sides to the point of standoff. The distinguishing military feature of the Cold War was what contemporaries called the 'balance of terror' (Galloway, 1972, p. 19). In the Western philosophical tradition the sublime has been associated with terror and the law of the father (see Giblett, 1996, especially chapter 2). The battlefield, for de Landa (1991, p. 82), is 'first and foremost a place of terror', or 'a zone of terror', as Barbusse (2004, p. 298) puts it in his 1916 classic of the First World War, *Under fire*, and as exemplified in Jünger's (2004, especially pp. 32, 71, 77, 215 and 225) 1920 classic of the First World War, *Storm of steel*. The circuit of association between the sublime, terror, war and the law of the father is complete.

The zone of terror was extended to new heights in subsequent wars. The launching of rockets by the Nazis late in the Second World War was not only, as Cornwell (2003, p. 145) puts it, 'a combination of advanced technology and fantasies of phallic terror', but also the extension of the battlefield, and the place of terror, to unmanned flight. The launching of Sputnik early in the Cold War extended the boundaries of the battlefield, and so the place of terror, to include orbital extraterrestrial space. It also paved the way for the post-Cold War imbalance or unbalance of terror in the age, the now, of terrorism.

After the foiled bombing of the World Trade Center in 1993 Virilio (2000b, p. 19) commented that 'after the age of the *balance of terror*, which lasted some forty years, the *age of imbalance* is upon us' as the bombing of the World Trade Center in 2001 attested. If the collapse of the Berlin Wall and the break-up of the Soviet Union in 1989 marked the end of the Cold War, the Gulf War of 1991 was the full birth of the *im-* or *un-*balance of terror culminating in 'the first NATO war' in Kosovo of 1999 (see Virilio, 2000b, p. 24; 2000c, pp. 5 and 58). In the post–Cold War era, terror has become unbalanced. The end of the Cold War did not mark the beginning of a new era of peace, but of a new type of terror as the bombing of the World Trade Center in 2001 indicated and its rejoinder in Iraq indicates.

With the satellite, orbital extraterrestrial space has ceased to be the heavens where the stars twinkle through the fourth window and has become a place of terror because it provides the military means of communication between points on, and surveillance of, the earth. Prominent here are 'remote sensing' satellites in synchronous orbit with the rotation of the earth 745 kilometres above the earth. They are used for what the United Nations Committee on the Peaceful Uses of Outer Spaces describes as 'identifying the nature and/or determining the conditions of objects on the Earth's surface, and of phenomena on, below, or above it' (cited in Mowlana, 1997, p. 119).

Even these 'peaceful uses', however, betray the military origins and display the earthly mastery of remote sensing. As Demac (1990, p. 214) argues, 'this technology was originally developed for military surveillance, to observe troop movements and to photograph military installations'. Even the 'peaceful uses' of satellites by foreign powers are a violation of national territorial rights. As Smythe (1981, p. 316) argues, 'the United States LANDSAT satellites invade the prerogatives of a country by sensing and mapping its mineral, crop and other physical characteristics without its permission'. Remote sensing is military mastery of the earth, whatever beneficial uses it may (and does) have (see Cubitt, 1998, pp. 45–55; Okolie, 1989, pp. 89–149).

The launch of Sputnik in 1957 ushered in the next decade of what Virilio (1994a, p. 201) calls 'the conquest of orbital extraterrestrial space', or what he earlier called 'the military conquest of space' (Virilio, 1986a, p. 26; 1991b, p. 23), or at least of what he calls 'the grand illusion of the so-called conquest of space' (Virilio, 1995, p. 109; see also Noble, 1999, pp. 115–142, especially p. 121). This conquest took what Smith calls 'the simulation of territorial and scientific conquest' (quoted in Wilson, 1992, p. 287) to new heights (literally) by exploring and crossing (but never closing unlike its terrestrial counterpart) the 'new high frontier' of space (in an extraterrestrial repetition of 'lighting out for the territories'), colonising orbital extraterrestrial space, enclosing the heavenly global commons and using it for terrestrial military purposes by taking 'the God's-eye view' from above to look down on what is below (see Cubitt, 1998, p. 48; Edwards, 1996, pp. 134–135). Subsequent decades have only extended and heightened this process not only into extraterrestrial space but also into cyberspace (as we shall see in a subsequent chapter), both of which colonise time and enclose the commons of the future (see Cubitt, 1998, p. 48).

The 1960s were the decade of the conquest of orbital extraterrestrial space; the 1980s were the decade of the militarisation of orbital extraterrestrial space (and the 1990s were the decade of paradoxically both the militarisation and the civilianisation of cyberspace (as we shall see in the next chapter)). In 1990 Hamelink (1990, p. 222) argued that 'outer space has become a militarized zone, since it is estimated that about 75% of satellites launched over the past decade have been for military use'. And used they were in the 1991 Gulf War where, and when, for Virilio (2002a, p. 4), '*the military satellite revolutionized the art of war between 1990 and 1991*' (his emphasis). By transcending the immanent, the satellite created an all-knowing, all-powerful, all-seeing eye and voice of the secularised gods of military might stationed on their 'modern-day Olympus' (Demac, 1986a, p. 36) of orbital extraterrestrial space. The 'artificial mountain tops' (O'Grady quoted in Moyal, 1984. p. 235) to which satellites supposedly gave access were, in fact, the slippery slopes of military might.

The civilian uses of communication satellites can never get away from their military beginnings. Unlike television (as we saw in the previous chapter), satellite technology, Raymond Williams (1989, p. 122) argues, was 'wholly dependent on major research, development and investment in a quite different field: that of military rocketry and its associated communication and espionage systems'. Even the civilian uses of satellites reproduce their military characteristics. Satellites transcend geographical space as they are 'distance-insensitive' (Hanson, 1982, p. 243; Cunningham, 1997, p. 100). Satellites, like the railway, telegraphy, photography and the car, not only overcame distance but also were insensitive to it and to terrain, whereas the railway, telegraphy, photography and the car were sensitive to the landscape to some extent, in terms of both how they travelled through it and how they viewed it.

Satellites are not only insensitive to distance but also, as Demac (1986a, p. 36) argues, 'insensitive to terrain and can penetrate deserts, jungles, mountains or Arctic outposts'. The reduction of distance, penetration of terrain (part of the way in which 'men' have 'known' metaphorically the ecosphere sexually, or violently) and the instantaneity of communication by satellites were all instrumental in producing McLuhan's idea of 'the global village', or 'the global supermarket' as Seddon (1997, p. 113) prefers to call it.

Early broadcast radio and television in the 1920s and 1930s exploited the resources of the electromagnetosphere to communicate to a mass audience (as we have seen in previous chapters). With the development and introduction of civilian communication satellites in the 1960s beginning with TELSTAR in 1962, 'the first true communication satellite' (Hanson, 1982, p. 246), broadcast radio and television exploited the resources of both orbital extraterrestrial space and the electromagnetosphere in order to communicate between many point sources and many more destinations over longer distances.

Satellites can thus be considered as an extension of radio; satellites took off from where radio left off. As Raymond Williams (1989, pp. 121–122) argues,

> the general usefulness of communications satellites had been hypothesized before they were practical. What was foreseen was an improvement comparable to that of early radio signals: that they could improve the signals from existing earth-based transmitters – whether in telephony or eventually in broadcasting – and they could take signals to many hitherto inaccessible places.

To do so, satellites, like their antecedent in radio, used the electromagnetic environment. Satellites transcend the terrestrial space of the inorganic, biospheric and atmospheric environments, but use the electromagnetic environment and orbit in extraterrestrial space. Communication satellites defy national boundaries for 'the satellite footprint marches boldly into neighbouring countries' (Demac, 1986b, p. xii) like a sort of *Blitzkrieg* (literally 'lightning war' (Fuller, 1943, p. 32; Virilio, 2002a, p. 142, n.2)) by Bigfoot Big Brother.

Following its militarisation in the 1980s, the 1990s may go down in history as the decade that opened up orbital extraterrestrial space as the new, decisive front in all terrestrial warfare. The Tofflers (1995, p. 78) commented after the 1991 Gulf War that 'the front was no longer where the main battle occurred'. Or more precisely, the terrestrial front was no longer where the main battle occurred as it had already been fought and won in space, and a new front, and frontier, opened up in space. This front was where the main battle had occurred. The Iraqis could never win the war on the ground, or in the air, or on or under the sea or in the electromagnetosphere because they had lost the war in extraterrestrial space before a shot had even been fired. To win the war on land, on or under the sea or in the air or in the electromagnetosphere, supremacy in orbital military space is now the determining factor.

Control and use of extraterrestrial space gave the United States Coalition forces a godlike vantage point, and advantage, to see and act, and the sublime power to destroy. For Virilio (2002a, p. 47) writing during the 1991 Gulf War,

> With satellite ubiquity and the instantaneity of military telecommunication, this overexposed war assumes the traditional attributes of the divine, so much so that on one side is the *mystical* fundamentalism and Saddam Hussein's call for a holy war, while on the side of the allies, we see a sort of technical fundamentalism, a call to pure war, with the support of sophisticated *materiel* (cruise missiles, smart bombs, etc.) that allows confrontation with the enemy almost without touching, *as if by nothing less than a miracle*, with the electromagnetic environment above the Iraqi territory effectively substituting for the normal milieu, the sphere of armed men.

Communication technologies are a secular theology of the sublime which allow their wielder to act through seeing, almost without touching and certainly without hearing and smelling. The cruise missile, for Virilio (and Lotringer, 2005, p. 79), is 'the emblem of the vision machine' as the latter was developed for the former (p. 78). The casualty figures on both sides testify to the imbalance of killing power. More than 150,000 Iraqi soldiers and civilians were, as Robins (1996b, p. 74) puts it, 'slaughtered in that bloody kill' compared to 300 Americans (cited in Robins, 1996b, p. 73), a ratio of 500:1. In probably no other war in history has the ratio of one side's casualties to the other's been so high. In the Iraq War begun in 2003, a much more conventional ground warfare of invasion and insurgency, thousands of Americans and many more Iraqis have been killed and wounded.

Satellites were the means to achieve supremacy in the 1991 Gulf War, though air supremacy was also gained by the introduction of a virus into the Iraqi Air Defence Command and Control System (Young and Jesser, 1997, p. 163). The Gulf War of 1991 was what Virilio (2002a, pp. 40, 44 and 46; his emphasis) calls the *'first total electronic war of history'*, not only because it was the first satellite war, but also because it was the first cyberspace war. Commenting on the 1991 Gulf War, Toma (1992, p. 2) argues that

> Satellites were the single most important factor that enabled U S land-based forces to transition quickly and smoothly from almost nothing to an extensive tactical communication network.

The United States could exercise satellite supremacy because it had space supremacy, both in extraterrestrial space and in cyberspace. Commenting also on the 1991 Gulf War, Brown (1996, p. 35) argues that

> Through the control of space, Allied forces had satellite information and imagery and a robust command, control and communications system; Saddam Hussein did not. We [*sic*] enjoyed access to our Global Positioning System or GPS; he did not. As such we could locate targets and hit them with timeliness and precision; Saddam could not.

One American journalist (cited in Tofflers, 1995, p. 119) commented that 'in the Gulf War we faced no attempts to blind or disable out satellites, and our enemy had no access to space for its own purposes'. Or more precisely, the Iraqis had access to civilian international satellite networks, as Anson and Cummings (1992, p. 122) argue, but they had 'no military space assets'. Coalition forces exploited this advantage to the full and demonstrated the new maxim of military and earthly mastery that 'who rules circumterrestrial space commands Planet Earth' (Tofflers, 1995, p. 124).

In the 1991 Gulf War against Iraq and in the 1999 NATO air strikes against the Serbs (albeit with some 'human error' and 'collateral damage' (see Virilio, 2000c)) US forces used communication satellites to fire missiles and drop bombs with deadly accuracy and devastating force. The 1991 Gulf War, for Virilio (1999, p. 99; 2000b, p. 25), 'was waged from the skies by satellites...that *deus ex machina* that manages the time of war'. Thus for him it was not only the first war to use satellites effectively but also 'the first war in real time' and 'a total war in which local space disappeared in a global and instantaneous military management operation'.

One military analyst enumerated some of the key communication features of the 1991 Gulf War:

> First, the coalition strategists became arguably the first wartime generation of strategists to have unbroken optical contact and other sensory contact with their enemy. Second, given the sophistication and definition of images provided by reconnaissance satellites, they had the confidence that they could 'believe their eyes'. Third, in view of the fact that virtually nothing could escape being sensed, what could be sensed could be targeted; and what could be targeted could be bombed, rocketed, strafed – in a word, killed. And fourth, since intelligence was available without interruption and in 'real time', a single composite was formed consisting of sensing, detecting, targeting and killing. Fused were the human eye, the satellite eye, the target and the selected weapon: it was the nearest example history has seen of the ultimate hand-eye co-ordination.
>
> (cited in Galvin, 1994, p. 178)

By being able to believe their eyes about what they saw in real time the coalition generals achieved 'a sort of continuous temporal outflanking', as *The Economist* put it. As a result 'Iraq's radar eyes were poked out, its wireless nerves severed' (cited in Robins and Webster, 1999, p. 158), and its electric body politic short-circuited. An Air Force official (cited in Campen, 1992a, p. xi) succinctly put it: 'we could see, hear and talk all through the war. After a few hours, he [Saddam Hussein] could not.' His body electric politic was disabled whereas the coalition's was enabled.

For the coalition generals the computer fused eye and hand, tool and weapon. For Edwards (1996, p. 14), computers are 'a primary example of the inseparability of weapon from tool'. Their functions were also inseparable, as Virilio (1989b, pp. 83 and 83) sums it up: 'nothing now distinguishes the functions of the weapon and the eye; the projectile's image and the image's projectile form a single composite...the eye's function being the function of a weapon'. Not only the gaze, but also the eye has been militarised. The eye is a weapon.

If cinema, and video, isn't 'I see', it's 'I fly', as Nam June Paik put it (quoted in Virilio, 1989b, p. 11; 1986a, p. 30; 1991b, p. 26), then satellites in war aren't 'I see', they're 'I kill.' The central concept of this new war game, for Virilio (1998, p. 167; 2002a, p. 110), becomes 'first look, first shot, first kill'. Satellites *are* artillery, or perhaps precisely the sighting and aiming mechanism for the 'artillery' of smart bombs whose modus operandi is, as a former US Under-Secretary of Defense described it, 'as soon as you can see it you can hope to destroy it' (quoted in Virilio, 1989b, p. 4; 1991b, p. 130; 1994b, p. 69; 1998, p. 144). With the rapid-firing gun, as we saw in a previous chapter, the function of the weapon is the function of the eye. With the precision-guided weapons (or PGWs) of the Vietnam War, Hanson (1982, p. 284) argues, 'a new military motto' was formulated: 'if you can see the target, you can blow up the target'. With the precision-guided munitions (PGMs) of the 1991 Gulf War the role of the eye and the weapon of the rapid-firing gun were reversed. During that conflict the function of eye is the function of the weapon.

This reversal was made possible by Tactical Mapping Systems (TMS), in which, as Virilio (1986a, p. 30; 1991b, p. 27) puts it, 'the functions of eyesight and weaponry melt into each other'. They also melt into the arm as, for Virilio (1991b, p. 130), 'the function of the eye becomes simultaneously that of the arm'. The arm is armed, arms are both weapons and limbs. God and the emperor had the power of the hand; 'man' not only has the gaze, as Barthes put it, but also the power of the hand, the arm and the weapon.

During the 1991 Gulf War the development and deployment, as Brown (1996, p. 31) outlines it, of

> 'Smart' weaponry – manifested in bombs, missiles, artillery projectiles and even infantry weapons systems – have created a new category of weapons called precision guided munitions (PGMs) and have led to an environment where 'everything that can be seen can be hit, and everything that can be hit can be killed'.

Continuous, space-based satellite and aircraft surveillance not only made the battlefield and the enemy visible day and night but also guided weaponry to its target turning 'dumb bombs' (Campen, 1992a, p. xi) into 'smart bombs'. Seeing the enemy meant that the soldier can (and will try to) kill him or her. Machines did the surveillance, sighting, targeting, firing, killing and wounding; the soldier merely pulled the trigger.

The satellite eye is a technological prosthesis for the human eye sublimated into a godlike eye in the sky. It is the eye of god in the sky, in extraterrestrial space, that looks and masters terrestrial space, and in war makes possible killing at the same time. With the rapid-firing gun, two human eyes became aligned with the weapon; with the satellite eye, no single pair of human eyes was aligned with the weapon but many human eyes. During the 1991 Gulf War, these eyes included the eyes of the generals and eyes of television viewers who saw the images transmitted from the nose cone of a smart bomb homing in on its target. During the live television coverage of the 1991 Gulf War Edwards (1995, p. 75) argues that 'as we rode the eye of the bomb we experienced at once the elation of technological power, the impotence and voyeurism of the passive TV audience, and the blurring of boundaries between "intelligent" weapon and political will'. In fact, the smart bomb was a weapon of political will that (dis)empowered the viewing voter: empowered to see, to know; disempowered to do anything, to move.

Not only the screen, as Robins and Levidow (1991, p. 325; 1995b, p. 106) argue, mediated the conflict to the point that 'the screen became the scene of the war', as they (1995b, p. 106; 1995a, p. 120) also put it, but also the satellite mediated the conflict and constructed the *mise en scene*. Desert storm was 'desert screen', as Virilio (1998; 2002a) puts it succinctly in the sense both that it was televised 'tele-action' and that 'the desert is a screen' against which the US coalition projected 'weapons of communication' and CNN projected 'weapons of mass communication' (Virilio, 2002a, pp. 7, 12 and 26; 2005b, pp. 32 and 87). For Robins (1994, p. 312), the 1991 Gulf War was 'a push-button, remote-control, screen-gazing war'. The satellite and the long shot displaced and distanced the combat from us at the same time as implicating us in it with their deadly accuracy. Television screened the desert screen. Just like the star, the desert became the surface against which masculine phantasy and power were projected and played out. Not only in the 1991 Gulf War, as Robins and Levidow (1991, p. 325) put it, did 'the vision of the long camera shot extend... the moral distanciation of previous wars' in which 'seeing was split off from feeling' (Robins and Levidow, 1995a, p. 121; 1995b, p. 107), but this vision and distanciation were mediated, made possible and reinforced by satellite communication and computers (see Edwards, 1996, pp. 354–356).

With the use of satellites in the 1991 Gulf War, orbital extraterrestrial space became the fifth front of military conflict after the four fronts of land, air, sea and the electromagnetosphere. Up until this war, all wars have been fought on land, and/or in the air and/or on or under the sea and/or in the electromagnetosphere (for example, by blocking radio transmissions or by decoding enemy secret radio transmissions). In the 1980s President Reagan wanted to militarise extraterrestrial space further with his Strategic Defense Initiative or 'Star Wars' technology by constructing a Maginot line of laser-firing satellites to protect the

United States against what Edwards (1987, p. 59) calls 'the terror of attack from the sky'. This initiative was later dropped (only later to be revived by President George W. Bush as 'National Missile Defense'), ostensibly because it was too expensive but this does not mean that orbital extraterrestrial space has not been, or had not already been, militarised. Indeed, the militarisation of orbital extraterrestrial space occurred with the launch of the first Sputnik in the Cold War.

The Gulf War of 1991 was the first war 'won' (though it was a curiously 'unaccomplished war', as Virilio (1994a, p. 203) calls it) in orbital extraterrestrial space. During this war Virilio wrote that 'the war in the Persian Gulf... will be seen to have inaugurated a radically new period in strategy characterised by *inertia*'. This includes 'the inertia of the centralised allied command, with all the techniques of instantaneous telecommunication at its disposal, precluding any other military movement than those of its own air force'. He draws the conclusion that 'victory in the air and especially victory in outer space' occurs 'for the first time in history' (Virilio, 1994a, p. 203). Telecommunications (especially via satellites) made possible this paradox of human bodily inertia and immense technological speed. This paradox was evident in the fact that, as Virilio (1994a, pp. 201–202) argues, 'the logistical problems of rapid transport and ultra-rapid communication have won out over the more traditional problems of battle and its tactical and strategic manoeuvres'. Satellite photography supersedes aerial photography; communication supersedes strategic warfare; rapidity supersedes manoeuvring.

The use of satellites in the 1991 Gulf War changed the nature of war forever by opening up a new front, the fifth front of orbital extraterrestrial space. Orbital military space, for Virilio (1994a, pp. 202 and 205; 2002a, pp. 3, 67–70),

> is now the determining factor in the critical parameters implied in winning land, sea, and air conflicts... What is now essential is over our heads, above the stratosphere, in that circumterrestrial void where a frightening number of unidentified flying objects [ufos] are orbiting, these arms of the fourth front that reigns over the three other three (sea, land, and air) – arms whose absolute power stems from the broadcast and reception of electromagnetic waves, radioelectric signals, and laser rays operating at the speed of light.

Elsewhere Virilio (1998, p. 174; 2000b, p. 21; 2002a, p. 121) calls this front 'the front of instantaneous electronic intelligence', 'the power of information' and the 'media front'.

In this respect Virilio is merely following in the footsteps of some high-ranking American military theorists. United States Air Force Colonel James McLendon (cited in Brown, 1996, p. 44) wrote that 'Information Warfare adds a fourth dimension of warfare to those of air, land and sea.' Furthermore, General Ronald Fogleman (cited in Brown, 1996, p. 44), former Chief of Staff of the United States Air Force, suggested that

> this information explosion ... signals that we're crossing a new frontier. Information has an ascending and transcending influence – for our society and our military forces. As such, I think it is appropriate to call information operations the fifth dimension of warfare. Dominating this information spectrum is going to be critical to military success in the future.

Brown (1996, p. 52, n.19) notes that 'Fogleman argues that space is the fourth dimension of warfare.' Space, for Anson and Cummings (1992, p. 121), 'added a fourth dimension to the war' in the Gulf. Controlling space is crucial for successful control of information as we shall see in a later chapter.

Terror over our heads

Rather than orbiting in the sublime company of heavenly bodies, satellites hover in the sublime place of terror. Arguably the electromagnetosphere is the fourth front after land, air and sea, whereas orbital extraterrestrial space is the fifth front with mastery over the first three using the resources of the fourth front. These two fronts are environments and are extensions of the other three. The first environment is the biosphere; the second, the inorganic environment; the third, the atmospheric environment; the fourth, the electromagnetic environment; and the fifth, orbital extraterrestrial space. The fifth environment of orbital extraterrestrial space is the fifth front of warfare, as was the case in the 1991 Gulf War. This front relies on the fourth environment of what Caldwell (1984, p. 234) calls 'the electromagnetic environment that surrounds the earth, permeates the atmosphere and is a phenomenon of the terrestrial planet itself'. Blocking radio transmissions in war and encrypting radio war communications show how this fourth environment became the fourth front of war. Orbital extraterrestrial space has become the fifth front in war and now a front in the global media wars; perhaps for the media it is the third front after developing terrestrial studios, transmitters and audiences, and gaining access to broadcast spectrum in the electromagnetosphere.

The fifth front, and environment, now has supremacy over the other four (satellites *are* artillery). Comparing the 1991 Gulf War to the Second World War, Virilio (1994a, p. 205; 2002a, p. 3; 2005b, p. 59) quotes General Heinz Guderian, the victor in the *Blitzkrieg* of 1940, who declared that 'Where the tanks are, that is where the front is.' General 'Stormin'' Norman Schwarzkopf was the victor in the military satellite communication big footprint *Blitzkrieg* of 1991. Virilio (1994a, p. 205; 2002a, p. 3) goes on to argue that

> Henceforth, this sentence [of Guderian] is definitively null and void, and has been replaced by this one: 'Where the satellites are, that is where the fourth [or fifth as I prefer] and final front is', the front of these arms of instantaneous intelligence and destruction that annul all military power.... The victory of the allied aerial forces in the Persian Gulf does not represent the end of land armies and the arrival of air supremacy, succeeding the time of the sea power of the battleships of yesteryear. Rather, it signifies the advent of an arms system whose power is literally exorbitant [see also Virilio 2002a, p. 4], in which the speed of communication and guidance of vectors of destruction's delivery from outer space annihilate all offensive capacity founded on movement, the assault of mechanised terrestrial forces.

The power of the communication arms system orbiting in extraterrestrial space is ex-orbitant in the sense of being outside the orb of the planet earth though it is still inside the orb of the ecosphere, of terrestrial phenomena, and orbits above it.

Supremacy in orbital military space can only be achieved by those who have access to the satellite technology to exploit the advantages of a strategic position in that new place of terror. If they do, they can exploit, as Virilio (1994a, p. 202; 1998, p. 180; 2000c, p. 14; 2002a, pp. 121 and 131; see also 2000b, pp. 21 and 25; 2001, p. 86) puts it, 'the supremacy of the *arms of communication* over *arms of massive destruction*' with 'the *vertical* dimension winning out...over the *horizontal*'. This vertical dimension not only refers to space, to height, but also to time, to synchronicity and to instantaneity. The fourth (or fifth, as I prefer) front, for Virilio (1998, p. 174; 2002a, p. 121), is 'a purely temporal dimension' in which 'real time...dominates real space'. The supremacy of the arms of communication amounts to the supremacy of the sublime over the slime and the supine, the vertical screen of stars over the horizontal window to the stars and the triumph of the nerds over the jocks, or more precisely to the fact that the nerds have given the jocks their most powerful weapon so far, sublime communication technologies. These technologies have given the generals the ability to see more piercingly into enemy territory, to know more profoundly, to fire more accurately, to penetrate more deeply, to destroy more ruthlessly.

The arms of communication have conquered terrestrial and orbital extraterrestrial space. In the light of subsequent history, Arendt may have been right in seeing the launch of the first satellite as a more momentous event than the splitting of the atom. The supremacy of the electronic arms of mass communication (satellites *are* artillery for bombarding audiences with messages and information) over the nuclear arms of mass destruction was achieved without the campaigners for nuclear disarmament and other peace activists noticing it because it was largely invisible or seemingly tangential.

The media have been more observant and astute, yet perhaps without appreciating the implications of what they were doing. *Asiaweek* once voted Rupert Murdoch the fifth most powerful person in Asia behind the leaders of China, Japan, Malaysia and Indonesia, and so even more powerful than the leaders of India and Pakistan, despite the display of their nations' nuclear capability. Why? Perhaps because Murdoch's News Corporation is a majority shareholder in satellite STAR TV with a footprint over most of Asia and with a potential audience of half the world's population (Man Chan, 1997, p. 94). Viacom CEO Sumner Redstone has accused Murdoch of wanting 'to conquer the world. And he seems to be doing it.' Ted Turner sees Murdoch in even more sinister terms as Adolf Hitler (cited in Herman and McChesney, 1997, p. 71). This analogy is perhaps not as bizarre as it might seem, and as mere commercial sour grapes, given that Murdoch may have been the victor in the civilian satellite communication big footprint *Blitzkrieg* of the 1990s.

The arms of communication do not just have a military connection or military use but are military by nature because they involve the conquest of space (both terrestrial and extraterrestrial) through their mastery of observation and surveillance, of communication and inform-ation, of knowledge and power. As Wilson (1992, p. 170) concludes, 'the low-intensity regional wars now favoured by modern states depend upon a system of command, control, communications [what in military parlance is called 'C3'] and intelligence that is probably the most complex organisation ever constructed'.

Ten years after the first launch of Sputnik, the Outer Space Treaty of 1967 declared outer space to be a commonly held resource not subject to national or private sector appropriation (Demac, 1986a, p. 37; 1986b, pp. xiv–xv). Orbital extraterrestrial space and the electromagnetosphere are no longer a global commons. They have been enclosed by the public/private sphere of civil society (such as corporations) and colonised by the public sphere (especially nation-states) in a typical act of co-operative resource exploitation of the ecosphere (for the spheres see Giblett, 2004, chapter 2). Enclosure of the commons into private property by modern corporations and colonisation of nature by states, and later nation-states, initially occurring on a terrestrial level from the sixteenth to the nineteenth centuries, has taken place on an extraterrestrial scale in the twentieth century (see Giblett, 1997a; Thompson, 1968, pp. 237–242).

When this shift took place precisely is difficult to determine, but it probably occurred in the late 1980s. In 1984 Caldwell (1984, p. 231) asked,

> will the outer space environment continue to be regarded as an international commons utilized by whatever nations have the technological capability to exploit it; will national jurisdictions be asserted, effectively as well as formally, upward and outward into space; or will the uses of space by managed be international agreement?

Extraterrestrial space has been conquered and colonised, enclosed and privatised, by co-operation between nation-states and global corporations with the resources to build, launch and buy functionality on satellites (see Demac, 1986a, p. 35; 1986b, p. xii). Yet national jurisdiction is not being asserted over extraterrestrial space in the same way as it was over terrestrial space. National boundaries do not apply in extraterrestrial space. Rather, corporate control is being exercised over the communicative space on satellites and over the entire ecosphere via vectors in the electromagnetic environment otherwise known as 'spectrum', a topic to which I return in the final chapter.

Modernity conquers, colonises and exploits the environment, not just the biospheric and inorganic environments, but also the electromagnetic and extraterrestrial environments. The last part of this process commenced around the time of the First World War, continued in the conquest of orbital extraterrestrial space in the 1960s and in its militarisation in the 1980s, and culminated in its terrestrial military use in the 1990s beginning with the 1991 Gulf War. Benjamin (1979, pp. 103–104; 1996, p. 486) noted in 1920s how, under the conditions of technologically advanced modernity beginning around the time of the First World War, human multitudes, gases, electrical forces were hurled into the open country, high frequency currents coursed through the landscape, new constellations rose in the sky, aerial space and ocean depths thundered with propellers, and everywhere sacrificial shafts were dug in Mother Earth. This immense wooing of the cosmos was enacted for the first time on a planetary scale, that is, in the spirit of technology.

With the launch of satellites and spaceships, this wooing took place on an extraterrestrial, even inter-planetary, scale. The boundaries of the technological mastery of nature have been extended from the terrestrial to the extraterrestrial.

This shift has profound implications for the role of technology in human affairs. Benjamin (1979, p. 104; 1996, p. 487) continues by bemoaning the fact that

> the mastery of nature, so the imperialists teach us, is the purpose of all technology. But who would trust a cane wielder who proclaimed the mastery of children by adults to be the purpose of education? Is not education above all the indispensable ordering of the relationship between generations and therefore mastery, if we are to use this term, of that relationship and not of children? And likewise technology is not the mastery of nature but of the relation between nature and man [*sic*].

Technology, in Buber's (1970, especially pp. 12–113) terms, is not (or should not be) an 'I–It' relationship to the world in which the world is objectified, 'I' am subjectified and 'I' exercise mastery over 'It' in a monologue, but an 'I–You' relation with the world in which the world is inter-subjectified and 'I' engage in dialogue and mutuality with 'You.'

By exercising mastery over nature (rather than over the relation between humans and nature), 'man' exercises mastery over humans as humans are part of nature. For Benjamin (1973b, p. 244; 2002, p. 121; 2003, p. 270), 'the destructiveness of war furnishes proof that society has not been mature enough to incorporate technology as its organ'. Technology remains outside the body of society as its tool rather than being brought inside to be an organ that functions in concert with the whole body. Benjamin (1973b, p. 244; 1999b, p. 312; 2002, p. 121; 2003, p. 270) argued that 'imperialistic war is a rebellion of technology' and predicted that 'any future war will also be a slave revolt on the part of technology'. The reason why all future wars have *not* been a slave revolt of technology is that technology is increasingly used to control technology; communication technologies are increasingly used to control armaments technology.

The fact that many communications technologies were developed for use in war and are used in war means that society has not been mature enough to incorporate them. War, as William Blake said, is 'energy enslaved'. As matter is energy, war enslaves matter too. As energy is a form of information, war is also now information enslaved. Norbert Wiener (1954/1989, p. 162), the 'father' of cybernetics, argued that 'the automatic machine...is the precise economic equivalent of slave labour'. War enslaves energy and information and the technologies of war put the slaves to work. War and technologies enslave not only energy and information but also humans. During the Vietnam War when computers were used to falsify information about the bombing of Cambodia, Admiral Thomas Moorer, Chairman of the Joint Chief of Staffs, complained that 'it is unfortunate that we had to become a slave to these damned computers' (cited in Weizenbaum, 1976, p. 239 and footnote).

Rather than 'man' relating to other humans and nature through technology, 'man' exercises mastery over other humans and nature to the point that, as Thoreau (1854/1997, p. 34) tersely put it in 1854, 'men [sic] have become the tools of their tools'. Such views were commonplace at that time for at about the same time Melville (1950, p. 221) inveighed against

> Machinery – that vaunted slave of humanity – here stood menially served by human beings, who served mutely and cringingly as the slave serves the Sultan. The girls did not so much seem accessory wheels to the general machinery as mere cogs to the wheels.

Melville juxtaposes this industrial hellhole of 'The Tartarus of Maids' to the pastoro-technical utopia of 'The Paradise of Bachelors'. The former makes possible the latter. With modernity 'men' have become the tools of their tools to the extent that Irigaray (1999, p. 147) asks, 'in this technical world he has fabricated, this world that resembles an organism that now has escaped him...has he not become a machine in the service of hi' creation?' The world has become a cyborg, and we are cyborgs who serve the cyborg of the world we have created.

The technologies of modernity colonised and enclosed the global commons of nature (including the human body and 'the environment'). The technologies of surveying marked out private property on the surface of the earth in the estates of the landed gentry and in the lots of

fragility and vulnerability.' Pen complements hand; hammer, arm; wheel, feet. Technologies, such as instruments, tools and transportation, *complement* deficiencies and extend physical capabilities, whereas communication technologies, such as the telegraph and computer, *supplement* physical capabilities and extend mental ones. Information technology sublimates physical abilities into mental capabilities, as Zuboff (1988, p. 23) puts it, 'by lifting knowledge out of the body's domain. The new technology signals the transposition of work activities to the abstract domain of information.' It also signals sublimation away from the concrete domain of energy and matter into the rarefied and ethereal sphere of binary data.

This process began with the telegraph by which, for McLuhan (1964, p. 248; see also p. 344), 'man [*sic*] had initiated that outering or extension of his central nervous system that is now approaching an extension of consciousness in satellite broadcasting'. No doubt, the computer and Internet constitute the further extension of consciousness following their precursors. The camera is a prosthesis for the eye and the mouth; the car a prosthesis for the feet; the radio, television; and the computer a prosthesis for the central nervous system. The computer is what Virilio (1991a, p. 48) calls 'an active, internal prosthesis of intelligence'. As this intelligence is then directed against the external to command and control it, the computer is an external prosthesis of belligerence. In previous chapters we have seen that the camera is a gun and the car a lethal weapon. Along these lines McLuhan (1964, p. 344) concludes that 'all technology can plausibly be regarded as weapons' just as all weapons are extensions of hands, nails and teeth for punching, gouging and crunching. Or perhaps in the case of computers, information technology is an extension of the fingers (though, of course, no fingers without hands). Fingers and nails, and even teeth, are digits, countable parts. Computers, according to Hafner and Lyon (1996, p. 27), 'had the potential to act as extensions of the whole human being' – body and mind.

Today's computers are digital, rather than analogue, microprocessors that reduce actions to events and inscribe them in binary code. Virilio (2000a, p. 2) critiques 'the decline of that *analogue* mental process, in favour of instrumental, *digital* procedures'. There is also the decline of physical processes in favour of mental ones, and of some physical ones over others in a process of sublimation. What Cubitt (1998, p. 151) calls

> The hierarchies of multimedia design have prioritized certain
> body elements – eye, ears, hands – over others, distracting and

disassembling the body in the interests of coherence now centred outside the body, in a pure communication between mind and object.

Following the decline of the power of the hand brought about by the eye, we now have the rise of power of the digit, the finger which is, as in Michelangelo's painting of God's creation of Adam, the power of God. By lifting a finger and pointing it to exercise power, 'Man' has the power of God. Power summons from a distance using the eye to see and locate the object and the finger to gesture and command. 'Man' exercises both the power of God and the gaze, hand and eye.

The computer as a creature of the mind is evident from its conception. Charles Babbage (1791–1871), according to Swade (1996, p. 34), is 'honoured as the towering patriarch in the history of computing' and as what Hanson (1982, p. 47) calls 'the true patron saint of the modern computer' (see also Shurkin, 1984, pp. 37–65; Campbell-Kelly and Aspray, 1996, pp. 10–15 and 53–60). Babbage is the monumental father of the computer, but it has no mother; it has a male saint, but no female one. It is, in short, a Bachelor Machine for a Bachelor Birth. Babbage (cited in Schaffer, 1996, p. 64) told Wellington in 1834 that his engines 'are the absolute creations of my own mind'. In Shurkin's (1984, p. 47) history of the computer, computers are 'engines of the mind' in general and Babbage's Difference Engine in particular 'substituted a machine for the human brain in performing an intellectual process'. For Swade, (1996, p. 47), Babbage's '1832 engine represents an ingression of machinery into psychology'.

Babbage conceived the idea of the computer, but he never gave birth to it. It had a long period of gestation of over a century until others gave birth to it. Roszak (1988, p. 19) argues that 'the first computer to enjoy a significant reputation was UNIVAC, the brainchild of John Mauchly and J. P. Eckery'. The computer, for Jastrow (cited in Roszak, 1988, p. 58), is 'the child of man's brain rather than his loins'. The brainchild is a Bachelor Birth from a Bachelor Machine. More recently, for Campbell-Kelly and Aspray (1996, p. 212), 'insofar as today's interactive style of computing can be said to have a single parent, that parent was [J.C.R.] Licklider', the MIT psychologist and computer scientist. Surely, and more precisely, Licklider was a single father whose labours gave birth, eventually, to the Internet, and not a single mother, that dreaded pariah of the media, that welfare cheat and parasite on the body politic, or corporate.

The computer had its beginnings in the military (see Edwards, 1987, pp. 45–60; Lévy, 2001, p. 13; van Creveld, 1991, p. 239). For Hafner and

Lyon (1996, p. 24), 'the relationship between the military and computer establishments began with the modern computer industry during World War II'. Yet the beginnings of the computer itself had strong ties with the military. The needs of war, for de Landa (1991, p. 129), 'have not only influenced the development of the internal components of computers (transistors and chips) but also computers themselves' right from their very beginnings. Babbage, according to Shurkin (1984, p. 47), believed that his invention 'would help with ballistics and navigation'. The primary purpose of Babbage's 'universal analytic engine' (designed but never built by him), according to Winston (1998, p. 156), was to produce nautical tables for the Navy. Winston (1998, p. 157) concludes that 'the creation of error-free mathematical tables thus had important military aspects'.

The Second World War was the watershed in the military development of the computer. In 1980 *Electronics* magazine concluded that 'weapons research for World War Two ... proved to be the catalyst in the creation of the electronic digital computer' (cited in Noble, 1984, p. 50). The transistor and the electronic computer, for Hanson (1982, pp. xii–xiii and 39), are 'by-products of war research' that 'emerged from the smoke and madness of World War II as the result of the explosion of classified research from 1940 to 1945'. Furthermore, for Hanson (1982, p. 61), 'the computer emerged from the war as a weapon and as a powerful, if still undeveloped, tool for mathematical logic'. For de Landa (1991, p. 41; see also pp. 42–3), 'the demand created by the military for cheap computing power ... motivated research in the automation of calculation'. UNIVAC, 'the first stored-program computer', was, Roszak (1988, p. 19) points out, 'based on military research done at the University of Pennsylvania during the [Second World] war'.

Not only was the conception of the computer couched in birthing metaphors (as we have seen), but also its emergence. For Noble (1984, p. 52), 'war-related developments in electronics, servomechanisms, and computers converged in the postwar period to create a powerful new technology and theory of control ... [as] scientists and engineers gave birth to a host of automatic devices and, most important, to a new way of thinking'. For Hanson (1982, p. 39) too, 'behind a wall [or screen] of secrecy, a remarkable aggregation of physicists, mathematicians and electronics experts gave birth to the computer age'. Yet 'fathering the unthinkable' of nuclear weaponry (Easlea, 1983) was, like all Bachelor Births from Bachelor Machines, masturbatory or 'onanistic', as Virilio (2002b, p. 20) puts it by quoting the physicist Freeman Dyson, who found it irresistible 'to feel it's there in your hands, to feel this energy that fuels the stars, to let it do your bidding'. Bachelor Machines are basically wanking machines. The first tool, the penis; the first machine, hand (or other wanking instrument) and penis – the first Bachelor Machine.

The first Bachelor Birth from a Bachelor Machine: the birth of Pallas Athene from the head of Zeus, an upward displacement and sublimation of the grotesque lower bodily stratum and the monstrous-feminine into the monumental-masculine. This birth is also what Virilio (2002b, p. 19) calls 'scientific integrism's irresistible mystical regression towards the Big Bang of the creation of the universe'. Sublimation of solid into gas involves a regression to the Big Bang, and a regression to the moment of conception, or at least of ejaculation, a big bang in another sense. In attempting to grasp the impossible object of the moment of one's conception, to think the sublime moment of one's coming into being, all one is left with is a Bachelor Birth from a Bachelor Machine.

Fathering the new way of thinking using information theory and fathering the computer age are Bachelor Machines for Bachelor Births, new brain-children and new worlds created without women and the Great Goddess of the earth. The computer, Winston (1998, p. 181) argues, that 'survived the Second World War *in utero* to be born as a child of the nuclear age', and of the Cold War. The computer brain-child was embryonic in the Second World War and born in the Cold War. The computer, Winston (1998, p. 321) also argues, emerged from 'Norbert Wiener's wartime work on predictive gun-sights' as what Roszak (1988, p. 202) calls 'his brainchild'. The computer is a war brainchild, the offspring of the war-bride machine. It still bears traces and scars of its parentage, as we all do. In the film and novel *Wargames* the computer is described as 'the child of war and as Wordsworth says, the child is the father of the man' (cited in Levidow and Robins, 1989b, p. 175).

Whilst the United Kingdom was constructing the welfare state in the aftermath of the Second World War, the United States was developing what Roszak (1988, pp. 41 and 203) calls the 'warfare state' with 'the government's continued heavy military investment in computers, electronics, and information theory following World War II'. Yet the United States had already become what Fuller (1946/1998, p. 209, n.3) called at that time the 'War State' with the approach of the Second World War. Indeed, as the United States was founded by war, it has always been a war state. The state of warfare has intensified to the point that the United States is now in a permanent state of alert. The perceived threat posed by the Soviet Union in the aftermath of the Second World War justified the formation and maintenance of what Noble (1984, p. 3) calls 'a permanent, global, peacetime military establishment'. Such oxymorons were the stock-in-trade of Cold War 'doublethink'.

Just as the satellite was the child of the Cold War as we saw in a previous chapter, so were its siblings, electronics and computers. The launch of Sputnik, Hafner and Lyon (1996, p. 20) argue, 'launched a golden era for military science and technology'. For Noble (1984, p. 7), 'the modern electronics industry...was largely a military creation'. Hanson (1982, p. 61) has described 'the computer as the automated central nervous system of military operations'.

The Cold War, Edwards (1996, p. ix) argues, 'shaped computer technology. Its politics became embedded in the machines...while the machines helped make possible its politics.' Cultural and politics forces shaped the technology, and those forces embedded in the machines further shaped cultural and politics outcomes in the feedback loop of the cultural determinism of technology (as we saw in the chapter on television). Winston (1998, p. 325) argues that 'the supervening necessity for networking the main frames came from the same military concerns as had caused those main frames to be built in the first instance... [thus demonstrating] the intimate connection of the computing project in general with the cold war and, specifically, nuclear confrontation' (see also Licklider, 1965, p. 17; Campbell-Kelly and Aspray, 1996, pp. 212–215; Hafner and Lyon, 1996, pp. 24–39; Hughes, 1998, pp. 255–265).

Similarly for Edwards (1995, p. 72; see also 1987, pp. 49, 51 and 53), 'there are strong, concrete connections between what I call the 'closed world' of post-WWII American global political hegemony and the 'microworlds' of computer simulations and artificial intelligence'. The microworld is in fact the tool that is used to try to control and command the (en)closed macroworld. There is always slippage and disjunction between the two. Fascination with and control of the micro and the minuscule does not lead to control of the functioning of the macroworld. Fetishism of the minuscule in 'toys for the boys' (both small and big) was a feature of both the nineteenth and the twentieth centuries that shows no sign of abating in twenty-first-century nanotechnology. The motto is 'control the big world by controlling the small'.

The alternative, for Edwards (1995, pp. 72–73), to this closed world is what Northrop Frye called the 'green world', 'an unbounded natural setting such as a forest, meadow, or glade'. The closed world is the fortress, the bunker, the modernist 'machine for living', the techno-pastoral utopia, whereas the green world is the pastoro-technical idyll of organic community. Both are products of the masculinist and monumentalist paradigm, as I have argued elsewhere (see Giblett, 2004). The alternative, and opposition, to both is the ecologically sustainable community living in bio- and psycho-symbiosis in and with its bioregion and the ecosphere, as I have also argued elsewhere (see Giblett, 2004).

The computer lived during the Second World War in embryonic form and emerged fully formed in the Cold War. Subsequent wars have been computer wars, at least for the United States. The Vietnam War was the first 'postmodern' or more precisely hypermodern war, the first electronics war and perhaps the first information war. The Vietnam War was, for Virilio (1989b, p. 82), 'the first electronics war in history' and the first television war transmitted to viewers at home (Young and Jesser, 1997, p. 79). For Jameson (cited in Giblett, 1996, p. 235), it was also 'the first terrible postmodern war'. Robins (1996c, p. 158) describes 'postmodern warfare' as 'increasingly a mediated affair, characterised by simulation, tele-presence and remote control'. Perhaps more precisely the Vietnam War was the first terrible hypermodern war as it was not postmodern at all, but excessively modern. Postmodern warfare is a contradiction in terms, an oxymoron. War went hyper with Vietnam whether it was hypermodern or hyperreal. For Bey (1996, p. 369), 'hyperreal war began in Vietnam with the involvement of television, and...reached full obscene revelation with the Gulf War of 1991'. The 1991 Gulf War for Virilio (1997a, p. 45) was 'the first "live" war', and for others (including Campen, 1992a and Robins and Webster, 1999, p. 157) 'the first Information War'. Both were also satellite wars.

Computers played a crucial role in the Vietnam War and all subsequent wars. Computers, for Gray (1989, p. 44), 'form the underlying basis and rationale for most post-modern war doctrines, policies and weapons, both materially and metaphorically'. The density of the rainforest canopy and the slimy swamps of Vietnam such as those of Mekong Delta and the Plain of Reeds (see Giblett, 1996, pp. 217–226), not to mention the perennial problem of seeing at night and the 'elusiveness of the enemy' (how inconsiderate of 'him'!), as General Westmoreland put it, created problems of surveillance whose solution was seen to lie in electronic sensors and information technology. The introduction and use of electronics and information technology had support at the very highest level. Westmoreland argued in this famous speech on 'the electronic battlefield' that in order to find and destroy the enemy it was necessary to fix the enemy and that to fix the enemy was 'a problem primarily in time rather than space' (see Dickson, 1976, p. 220; see also Littauer and Uphoff, 1972, pp. 151–166). It was relatively easy to locate the enemy in space, but not in time: would 'he' (it was always a 'he' for Westmoreland) still be there when one arrived? The location of the enemy had to be fixed in both space and time. The solution to the problem was seen to be located in time-fixing, and almost instantaneous, communication technologies. Westmoreland goes on to argue that 'on the battlefield of the future, enemy forces will be located, tracked, and targeted almost simultaneously through the use of data links, computer assisted intelligence evaluation, and automated fire control'.

The 1991 Gulf War was the precise fulfilment of this prediction as it was, for Campen (1992b, p. 135), 'an information war'. During this war, for Campen (1992a, p. xi), 'an ounce of silicon in a computer may have had more effect than a ton of depleted uranium'. The silicon had more devastating effects on the Iraqis and less harmful effects on the US personnel, whereas the depleted uranium had a damaging effect on Iraq and on the health of US soldiers who handled depleted uranium shells.

Westmoreland went on in what can only be described as a gross militaristic and parodic travesty of Martin Luther King's famous 'I have a dream' speech (see Dickson, 1976, pp. 71–2) to

foresee a new battlefield array. I see battlefields or combat areas that are under 24 hour real or near real time surveillance of all types. I see battlefields on which we can destroy anything we locate through instant communications and the almost instantaneous application of highly lethal firepower. (Dickson, 1976, p. 221)

The 1991 Gulf War was the realisation of Westmoreland's terrifying vision.

The Vietnam War produced what van Creveld (1985, pp. 246–9, 258–260) calls 'information pathology', an 'insatiable demand for information'. This demand was met in the Vietnam War to some extent by communications and data processing systems, especially Project Igloo White (see Dickson, 1976, pp. 83–99; Littauer and Uphoff, 1972, p. 151) with its electronic sensors and its 'gigantic computerized nerve center' in Thailand that ran from 1969 to 1972. Yet this was not a totally novel innovation as de Landa (1991, p. 79) relates that since the beginning of the World Wide Military Command and Control System in 1962 'computers slowly became the main instrument for the centralization of command networks'.

In what Weizenbaum (1976, pp. 238 and 239; my emphasis) calls

the war *against* Vietnam computers operated by officers who had not the slightest idea of what went on inside their machines effectively

chose which hamlets were to be bombed and what zones had a suffi-
cient density of Viet Cong to be 'legitimately' declared free-fire zones,
that is large geographical areas in which pilots had the 'right' to kill
every living thing ... In modern warfare it is common for the soldier,
say the bomber pilot, to operate at enormous psychological distance
from his victims. He is not responsible for burned children because
he never sees their village, his bombs, and certainly not the flaming
children themselves.

These US officers were just like those car drivers who have no idea how
their car works. This ignorance produced a similar sense of psychological distance and the impossibility of empathy, let alone compassion,
made possible by the physical distancing brought about by modern
communications technologies. For Dickson (1976, p. 88), some of the
bizarre missions run as a result of this surveillance and targeting 'must
be considered a milestone in the history of war in that technology now
permits one to listen in on an attack you are directing from a control
center hundreds of miles away. Only a TV picture is lacking.' The 1991
Gulf War was to make up for that deficiency by providing TV pictures
from the nose cones of 'smart bombs'. What Dickson (1976, p. 88) calls
'the blind bombing by computer' in the Vietnam War was to become
what would be called 'the seeing bombing by computer' of the Gulf War.

Information

Unlike the Second World War, in which information was regarded *as*
a weapon (see Edwards, 1987, p. 46), information from the Vietnam
War on *is* a weapon. Information is crucial for war today to the point
that warfare is information warfare. The concept and activity of information warfare brings together two powerful words yet whilst the latter
may not be contentious, the former is either taken for granted or, if
not, becomes a controversial topic. In information warfare, information is both 'weapon and target' (Hutchinson and Warren, 2001, p. 1).
Information, for Hutchinson and Warren (2001, p. 1), is 'collated data
in context'. Yet information is also commodified data in context.

Not only has information emerged recently as a means and object
of warfare, as Hutchinson and Warren argue, but it also, as Beniger
(1986, p. vi) argues, 'has only recently emerged as a distinct and critical
commodity'. If land was the economic base of pre-capitalist modes of
production and labour the base of the capitalist mode of production with
the Industrial Revolution marking the turning point between the two,

then what sort of revolution Beniger wonders will mark the transition to the 'Information Society'? The answer for Beniger is what he calls the 'Control Revolution'. He goes on to argue that

> the Information Society... is not so much the result of any recent social changes as of increases begun more than a century ago in the speed of material processing. Microprocessor and computer technologies, contrary to currently fashionable opinion, are not new forces only recently unleashed upon an unprepared society, but merely the latest installment in the continuing development of the Control Revolution.

Both the 'Control Revolution' and the 'Information Society' are further episodes in the long story of the bourgeois revolution and industrial capitalism. Just as 'land' or the earth still continues to be the economic foundation for all wealth (see Giblett, 2004, chapter 2), so the 'Control Revolution' still continues the bourgeois revolution of capitalism. History does not move in discrete stages in which each new stage supersedes the previous one. It moves by preservation and subjugation in which all previous 'stages' are subsumed in proceeding ones and in which traces of preceding ones still persist. The 'Control Revolution' is merely the latest instalment in the bourgeois revolution.

It is based on information theory which, Winston (1998, p. 322) argues, 'commoditises information, draining it of semantic content' and reducing it to packets of data that can be transported along cables, lines and vectors through the electromagnetosphere and extraterrestrial space. Rather than the dawning of a brand new day of freedom, the Internet represents just another day in the business of commodifying information. The Internet, for Winston (1998, p. 335), 'represents the final disastrous application of the concept of the commoditisation of information in the second half of the twentieth century' which shows no signs of abating in the first half of the twenty-first. Winston (1998, p. 336) concludes that 'the Information Highway will transform itself even more than it is at present into the Information Toll Road'. Only those rich enough to pay the toll will be able to drive on the highway.

The bourgeois revolution beginning in the late eighteenth century was characterised by the development of industrial capitalism and it was underpinned and made possible by enclosure of commons, initially land and water, later air, extraterrestrial space and the electromagnetosphere. Not only was, as Robins and Webster (1999, p. 7) put it, 'common land privatised' but also space (including orbital extraterrestrial space as we saw in the previous chapter) and spectrum (as well see in the next chapter). Robins and Webster (1999, p. 7) argue that 'the global network society ... promises to enclose the entire globe, creating what Stephen Gill has called the "global panopticon" '. They go on to argue that

> the new communication and information technologies – particular in the network society – permit a massive extension and trans-formation of that same (relative, technological) mobilisation to which Bentham's Panoptic principle aspired. What these technologies support, in fact, is the same mechanism of power and control, but now freed from the architectural constraints of Bentham's stone and brick prototype. On the basis of the information revolution, not just the prison or the factory, but the social totality, may come to function as the hierarchical and disciplinary machine.
>
> (Robins and Webster, 1999, p. 120)

Zuboff (1988, pp. 319–324) was perhaps the first to apply Bentham's Panopticon (1787/1962) and Foucault's (1977) work on it to information technology. After reprising both she goes on in a discussion of what she calls 'the panoptic power of information technology' to argue that

> Information technology ... can provide the computer age version of universal transparency with a degree of illumination that would have exceeded Bentham's most outlandish fantasies. Such systems can become information panopticons ... freed from the constraint of space and time.
>
> (Zuboff, 1988, p. 322)

Information technology combined the freedom from space (local place) that the railway gave and the freedom from time (local time) that the telegraph afforded. It also freed power from the constraints of the body. Rather than an apparatus for disciplining and regulating the body, Bentham (1787/1962, p. 39) proclaimed the Panopticon as 'a new mode of obtaining power of mind over mind, in a quantity hitherto without example'. By obtaining power over the mind, and by splitting mind from, and privileging it over, body, one could exercise power over the body.

Along similar lines, de Landa (1991, p. 205), drawing on Bentham's Panopticon and Foucault's work on it, develops the idea of 'the 'Panspec-tron', as one may call the new non-optical intelligence-acquisition machine. The Panspectron, or global panopticon, represents, for Robins and Webster (1999, p. 7), 'the forward march of the Enclosure movements'. This forward march is occurring just as much, if not more, in time as in space, foreclosing the future, the temporal analogue and counterpart to enclosing the commons. In a section entitled 'Enclosing the future' Robins and Webster (1999, pp. 232–237) argue that

> If the world's space has been colonised by the logic of order and rationalisation, so too has its time. A major achievement of the capitalist imaginary has been the colonisation [and enclosure] of the future – and that means the colonisation [and enclosure] of possibility... The future is no longer the other of the present – it therefore no longer contains the possibility of unknown encounters and events that would be transformative.

The future becomes the repetition of the present, and the present the repetition of the past. The history of the future entails both the colonisation of time and a repetition of the history of the past. Unlike the enclosure movement of the seventeenth and eighteenth centuries that enclosed common land in private property, the enclosure movement of the late twentieth and early twenty-first centuries is enclosing time.

The computer is the descendant of other communication and transportation technologies. The monitor is not only the inheritor of the television and cinema screen, but also of the train window and car windscreen. Virilio (2000a, p. 16; see also 1991a, p. 61) argues that 'the aim is to make the computer screen the ultimate window... Here the computer is... an automatic vision machine'. It is no accident that Microsoft's platform software is called 'Windows', the inheritor of the vertical railway and car window through which a motionless spectator views an active world while sitting now inside a vehicle that does not move, a sedan chair going nowhere.

The computer is also the direct descendant of the telegraph as it enables one, in the words of Licklider (1965, p. 6), to 'transmit information without transporting material'. In 1960 he envisaged '10 or 15 years hence, a 'thinking center' that will incorporate the functions of present-day libraries' and that there would be a 'a network of such centers, connected to one another by wide-band communication lines and to individual users by leased-wire services' (Licklider, 1960, p. 8). In 1968 he and Robert Taylor (1968, p. 37) discussed 'on-line interactive communities'. In 1965 Licklider (p. 33) predicted cautiously that 'by the year 2000, information and knowledge may be as important as mobility'.

The aim, for Licklider (1965, p. 32), 'is to get the user of the fund of knowledge into something more nearly like an executive's or commander's position'. In other words, the aim is to get the user into a position to command (give orders) and execute (orders, lives and so on), a position of military mas/tery. Sitting in front of the computer monitor giving commands on a keyboard or with a mouse is a military position of command, control and communication. This master of time and space (and slave to distance and haste) will drive 'his intellectual Ford of Cadillac' (p. 33) on the information superhighway. The dashboard is 'a display-and-control system in a telecommunication-telecomputation system' connected by the 'umbilical cord' of the cable by which cyborgs are connected to the mother ship. Information technology, as Levidow and Robins (1989a, p. 9) put it, 'bears the stamp of its military origin – not just as hardware or software, but as 'liveware', as 'man/machine interfaces', as social relations'.

Like the computer, indeed as the linking of computers, the Internet has military origins (see de Landa, 1991, pp. 117 and 120; Hauben and Hauben, 1997, pp. 41–2,49, 96–7, 116–118; Hughes, 1998, pp. 255–265). Yet the idea of the Internet has civilian origins as Jules Verne in 1889 had a prescient vision of the online newspaper. He described how, in the twenty-ninth century,

> in addition to his telephone, each reporter has in front of him a series of commutators, which allow him to get into communication with this or that telephotic line. Thus the subscribers have not only the story but also the sight of these events.
>
> (1889/1999, p. 195)

The Internet, as Virilio (2000d, p. 77; 2000a, p. 12) reminds us, is 'of military origins and has military purposes' but as 'the recently *civilianised* military network' it is returning to Verne's idea.

The civilianisation of military technology was both a product of and a resistance to the militarisation of civilian life that characterises modernity and increasingly characterises hypermodernity. The Internet, for Uglow (1996, p. 9), 'to begin with, seemed liked power to the people at last, an eager civilian takeover of military, state and corporate encampments'. It also entailed a takeover of computers and a rethinking of their functions with what Streeter (2000, p. 140) calls 'computers as devices for individualised symbol manipulation and horizontal communication, instead of as calculating machines or devices for control-at-a-distance'.

The Internet, for Streeter (2000, p. 131; see also Lévy, 2001, p. 208), is 'a decentralised communication system with a strikingly libertarian ethos was created within a military-oriented research and design enterprise famous for its hierarchical, authoritarian culture and organization'. The Internet is the paradoxical machine of hypermodernity par excellence, especially when its role in commerce is considered as well. For Lévy (2001, p. xi), 'there is no reason to separate commerce from the libertarian and communitarian dynamic that presided at the birth of the Internet. The two are complementary' in what he calls 'cyberculture'. Whether they are equal partners and can remain so is another question. The Internet may have shaken off its military mantle, but whether the libertarian and communitarian can continue to coexist with its commercial partner is another matter. The commercial may have replaced the military as partner of the communitarian and libertarian in 'Internet Inc.'

Whether the 'Inc.' stands for transnational corporations or for non-government organisations is an unanswered question, as is whether the non-commercial uses of the Internet are 'a new social movement' of a counterculture or 'the cyberculture movement', as Lévy (2001, pp. 14 and 209) claims; that is, whether cyberculture is countercultural. A distinctive and crucial feature of the Internet, and a vital departure from previous communication technologies, is the fact that, as Lévy (2001, pp. 44 and 65) points out, it communicates in what he calls a 'multilogue' from the many to the many unlike the one-to-one dialogic communication of telegraphy or telephone and the one-to-many monologic communication of cinema or of broadcast radio or television. In this respect the Internet is more like photography, where family and friends exchange photographs only on a grander scale where many can share their pictures and stories with many more.

Can the Internet shake off its military beginnings? Streeter (2000, p. 132) claims that 'a technological system born in the heart of the military–industrial complex came to embody distinctly non-military values'. A Bachelor Birth from a Bachelor military Machine grew up to disown its own father and to espouse civilian values. A machine that was born clothed in camo grew up to wear civvies. Yet the child could never change its spots given to it by its parents. The civvies over the camo were vestamentary testament to hypermodern militiarisation of civilian life.

The Internet Streeter (2000, p. 144) 'was largely born within non-profit institutions' such as research and educational institutions like universities supportive of the military, rather than in non-profit, non-government, voluntary organisations often opposed to the military. Universities were recruited to the ranks of what Streeter (2000, p. 145) calls 'the larger corporate framework wherein those non-profit institutions were understood to be supportive of the for-profit, corporate world'. The Internet, for Streeter (2000, p. 132), 'is not a utopian blueprint come alive, but what people have imagined the Internet to be is intimately and necessarily bound up with what it has become'. What it has been imagined to be is intimately tied up with its what it has been. It may deny, but it can never escape, its military conception.

Cyberspace Paradiso

The Internet is more than just the networking of computers. It produces cyberspace, a virtual place of interactions and imaginings. Cyberspace, for Fisher (1997, p. 114; see also pp. 122 and 125), 'is the post-modern paradise'. Arguably cyberspace is not postmodern enough, not *post*modern at all, but hypermodern. It is a capitalist market of ideas, and goods. Cyberspace, for Lévy (2001, p. 208), is 'big business'. Cyberspace, for Stone (1996, p. 33), is 'a space of pure communication, the free market of symbolic exchange'. Just as so-called 'free trade' is not fair trade, so cyberspace is not a fair market of symbolic exchange. Lévy (2001, p. 74; his emphasis) defines cyberspace as 'the *communications space made accessible through the global interconnection of computers and computer memories*'. Cyberspace, for Wertheim (1999, p. 258), 'could be a digital version of the [biblical] Heavenly City' as it is in Gibson's *Neuromancer* with its 'idealised polis of crystalline order and mathematical rigor'. In an interview on cyberspace Virilio (1997a, p. 44) claims that 'there is something divine in this new technology. The research on cyberspace is a quest for God. To be God.' Fisher (1997, p. 122) describes 'the synthesis of theological and technological discourses into a technosophy of cyberspace'.

As God created the world, so cyberspace creates a world, even a living being, such as when Dyens (2001, p. 30) defines cyberspace as 'a scattered, diffused, ever-changing organism, one that is just as alive as a wetland'. Cyberspace is to Lovelock's Gaia ('for whom the planet is a cybernetic organism' (Dyens, 2001, p. 49; see Giblett, 1997b and 2004, chapter 2) for a critique) as cyborg is to hypermodern human being. Yet cyberspace and wetland are qualitatively different as they perform different functions. Cyberspace is a sublime communication technology that transforms solid matter into gas, whereas the wetland is a slimy body of water and earth that transforms solid matter into liquid (see Giblett, 1996, chapter 2). The wetland also produces new living biological organisms out of dead and dying matter, whereas cyberspace cannot produce new life itself but produces new technological organisms (cyborgs) out of already living and dead matter (see Giblett, forthcoming).

'Cyberspace' is defined famously in Gibson's (1984, p. 51) novel as

> a consensual hallucination experienced daily by billions of legitimate operators...A graphic representation of data abstracted from the bank of every computer in the human system...Lines of light ranged in the non-space of the mind, cluster and constellations of data

producing 'the nonspace [and non-time] of the matrix' (Gibson, 1984, p. 63). Both the mind and the matrix are non-spatial and non-temporal in the Cartesian model of cyberspace. As Olsen (1991, p. 283) reminds us, 'the word *matrix* derives from the Latin for *womb,* which in turn derives from the Latin for *mother'*. Olsen draws the dubious conclusion that 'while it is true that only males have access to cyberspace, it is equally true that what they have access to is a female region'. Or more precisely, the matrix is a feminised region, a simulated womb to which men only (can) have access to in patriarchal heterosexuality. Robins (1996a, p. 10; 1996b, p. 91) critiques this 'idealisation of the electronic matrix as a facilitating and containing environment' when it is at best a technological analogue and at worst a mathematical projection (see Giblett, 1996, figure 1, p. 26).

The matrix is the grid-plan town, dead matter, empty or Jupiter (Jovean) space, the landscape of trench warfare, a lunar landscape, the end product of sublimation, the sublimate (also see Giblett, 1996, figure 1, p. 26). Matrix is not mater. Neuromancy is 'necromancy' by another name, by the slip of one letter, by a slip of the pen. By casting what Olsen (1991, p. 284) calls 'a mystical aura' around the matrix Gibson creates 'a cybernetic sublime'. Yet this aura is not the strange weave of space and time that Benjamin saw in early photographs, nor the aura of the distance from the unique appearance of an object in time and space as he also saw, but the transcendence of space and time into a spaceless and timeless void in which, as Virilio (2000a, p. 9) puts it, 'we are not seeing an "end of history", but we are seeing an end of geography' in the sense that we are seeing an end of a sense of local place, though not, of course, of global space. Quite the contrary. Geopolitics – power struggles over the earth and its space, and time (and across time-zones) – supersedes geography – the writing of the earth and its places.

The ghost of Descartes haunts computers and cyberspace. For Stone (1996, p. 34), 'the geometry of cyberspace as Gibson described it was Cartesian'. Yet, for Winston (1998, p. 148), 'the computer offends, fundamentally, the Cartesian duopoly of mind and matter'. Mind and matter merge, but mind remains pre-eminent over matter, including the body. The computer does not offend the Cartesian dualism of mind and body and the Cartesian monopoly of mind over matter. It is both 'an unthinkable instrument' and a machine that 'thinks'. For Licklider (1965, p. 91), 'man–computer interaction', even symbiosis (Licklider, 1960, p. 2), entails 'synergic action in which men and machine participate together'. In what? For what? Do mind and body merge? No, body is meat, so much living matter to be sloughed off.

In Gibson's (1984, p. 5) *Neuromancer* the matrix is 'bright lattices of logic unfolding across that colorless void' and 'the consensual hallucination' into which Case could project his 'disembodied consciousness' by jacking into 'a custom cyberspace deck'. Case 'lived for the bodiless exultation of cyberspace' and adopted 'the elite stance that involved a certain relaxed contempt for the flesh. The body was meat', and flesh a prison (Gibson, 1984, p. 6). The body, as Balsamo (1995, p. 229) puts it, is 'nothing more than excess baggage for the cyberspace traveller'. The computer itself may offend the Cartesian duopoly of mind and matter, but cyberspace reproduces Cartesian geometry and mind–body dualism. For Stone (1996, p. 34), 'cyberspace as Gibson described it was a physically inhabitable, electronically generated alternate reality…inhabited by refigured human "persons" separated from their physical bodies which were parked in "normal" space'. Fisher (1997, p. 113) remarks on the 'transcendence of the body and things bodily' and 'the sublation of the body' in 'the hypercorporeality of cyberspace'.

Yet cyberspace is not just a space but also a time. It is a set of space–time coordinates. Cyberspace, for Wertheim (1999, p. 263), 'becomes…a place *outside* space and time'. Cyberspace is sublime, a landless scape and timeless time. Landscape colonised space (see Giblett, 2004); cyberspace supersedes landscape to colonise space and time. Cyberspace colonises the matrix of time and space. As landscape is surface (of the land, of the landscape painting (see Giblett, 2004)) cyberspace is surface (of the monitor even with its 3D graphics); cyberspace is cyberscape. Telegraphy and radio colonise space and time; cyberspace transcends them. For Olson (1991, p. 283), Gibson's computer hackers 'travel…from the realm of *chronos* to the realm of *kairos*, from a materialistic geography registering realistic chronology, logic, and stability, to an ethereal one registering spiritual timelessness, alogic and possibility'.

The console cowboys travel from the mundane duration of everyday life rooted in local place to the divine timelessness of infinity that transcends time and space. Computer networks 'allow us to transcend both time and space' (cited by Gleick, 2002, p. 60) because we travel virtually through space, in almost real time. In a word, we are in the sublime. For Voller (1993, p. 18), 'the concept of cyberspace' central to William Gibson's 'matrix novels and stories is 'an extension of and comment upon one of the most significant elements of Romantic aesthetics, the sublime'. Central to the sublime, for Voller, is its search for 'intimations of the divine' and, I would add, 'intimations of immortality' as in the title of one of Wordsworth's poems, though divinity and immortality largely amount to the same thing in the Romantic lexicon. The divine is immortal; immortality is divine. The sublime is a timeless and spaceless realm.

Cyberspace, like space, is expanding. Cyberspace, for Wertheim (1999, pp. 224, 225), was 'expanding exponentially' in the 1990s, whereas in the 1970s its 'growth was necessarily incremental'. Cyberspace is incrementing exponentially; in a word, excrementally. The matrix is not just a virtual space and a consensual hallucination for hackers, console jockeys, email users and web surfers. For Uglow (1996, p. 150), 'our skies are a moving web of communications, planes and satellites'. Our skies, including our atmosphere, our electromagnetosphere and our orbital extraterrestrial space, are a moving web of communication and transportational vectors. Cyberspace not only, as Stone (1996, pp. 34 and 35) puts it, 'exist[s] now as a metaphor for late-twentieth [and early twentieth-first] century communications technologies', but also exists as a matrix of communication and transportational vectors. This matrix is what Virilio (2000a, p. 7) calls 'the electronic ether of our modern means of communication'. After the ether of the electromagnetosphere in radio comes the ether of telecommunication in the electromagnetosphere and orbital extraterrestrial space in television, telephony and the Internet.

11
Blue Sky Mining: Spectrum and Space

Electromagnetic spectrum for telecommunications and extraterrestrial space for communication satellites are two of the leading edge contexts for communication technologies today. Both have been colonised by nation-states and exploited by transnational corporations, yet both are part of the earthly household. Both spectrum and space have been debated as either property open for 'blue sky mining' or for preserving as a commons owned by none and shared by all. Whilst the former is business-as-usual in the heavens as on earth, the latter seems like pie in the sky, though zoning spectrum and space could ensure more equitable access to it and more equitable distribution of its benefits.

Electromagnetic spectrum

Mining the skies in a number of senses has been in the news over the past decade ranging from proposals for mining in space and so gaining 'untold riches' from the asteroids, comets and planets (Grossman, 2001, p. 35; Lewis, 1996) to selling off electromagnetic spectrum. Various frequency ranges associated with specific wavelength ranges are used, and suited, for particular radio applications (see Levin, 1971, Table 1, p. 21; Taylor, 1975, table 3, p. 211). Spectrum is variously regarded as 'a natural resource' (albeit invisible) (Levin, 1971, p. 1) to be exploited like any other natural resource; as 'a scarce resource' (Gleick, 2002, p. 45) administered by governmental and international agencies; as *not* a scarce resource which has only been constructed as such as digitisation eradicates scarcity (Jassem, 1998); as a common source of free goods to be colonised by nation-states and enclosed in the private property of corporations; and as a commons to be owned by none and shared by all who should equally enjoy its benefits.

Reports in *The Australian* newspaper a few years ago described the auctions in Australia for licences to use sections of the electromagnetic spectrum. One of the articles entitled 'Reach for the Sky' borrowed its title from Douglas Bader and his best-selling boy's own adventure story of flying Spitfires in the Second World War culminating in the Battle of Britain. Spectrum replaces Spitfires in this new war of the worlds, or the war against the world, that sees government and business locked into auctioning off, and bidding for, the global communication commons of the electromagnetosphere. The byline for this story reads that Federal Treasurer 'Peter Costello hopes to get billions for chunks of the wild [*sic*, not wide, or wired] blue yonder. Geoff Elliott asks whether he will be over the moon with the result' (Elliott, 2000, p. 21). More to the point, will the citizens of Australia be over the moon with the result? What involvement have citizens had in the sale? Will it benefit all equally?

Spectrum is figured here as the new wilderness with the governmental carving up of it as the new frontier and governmental selling of it as the new Eldorado with its 'glittering celestial promise' of 'glittering success'. No longer do we believe in a promised land, an earthly utopia; we now believe in a promised sky, a heavenly utopia of a networked world. No longer do we look for glittering goodies beneath the surface of the earth (see Giblett, 2004, chapter 9); we now look for glittering goodies in the heights above the earth, in the sky, in the electromagnetosphere, in spectrum. Whilst satellites merely orbit in the sublime company of heavenly body, spectrum holds out the promise of gold-mining the heavens. Satellites were merely the first step in 'man's' escaping the prison of the earth, whereas spectrum, for Elliott (2000, pp. 21 and 28), has seen 'telco and internet executives disconnect from the real world' and orbit in the 'rarefied' extraterrestrial and sublimated world of 'stellar valuations' where 'share prices soar'. And fall, if not plummet like Icarus, who flew too close to the sun and got burnt.

Costello may be over the moon and telcos may be 'all but promising a trip to the moon', yet the moon, like a mined landscape, is a dead landscape. The trench warfare of the First World War, the German Romantic love of nature and the German Idealist transcendence of nature produced a dead, lunar landscape as we saw in a previous chapter. I would not wish to be over, or to take a trip to, that moon. The depths and surfaces of the earth have been and still are mined, and now the extraterrestrial heights of the earth are being mined in spectrum sell-offs. Orbital extraterrestrial space was colonised in the 1960s by powerful and wealthy nation-states using rockets and satellites. After colonisation comes mining; Columbus followed by Cortez (see Giblett, 2004, chapter 9);

John Glenn by Bill Gates. The new world of the Americas gives way to what Elliott (2000, p. 28) calls 'a new world of communications'.

Elliott (2000, p. 21) draws a parallel between spectrum selling and 'what the stock promoters during the Poseidon mining boom of the late 1960s and early 1970s called Blue Sky Mining'. Electromagnetic spectrum is now being mined by blue sky miners that repeat the same gung-ho, go-get-'em, clichés as their counterparts in the 1970s. The only problem with this metaphor is that it, like many metaphors, has a sting in its tail. Blue sky mining also refers to the devastating blue asbestos mining as depicted in the protest song of the same title by the Australian rock music group Midnight Oil. Many miners and their families at the Wittenoom blue asbestos mine in Western Australia contracted asbestosis and later led a painful life and died a painful death. Blue sky mining can be bad for your health! And for the health of the land, the earth, for eco-health.

If, as Elliott (2000, p. 28) puts it, 'the world remains transfixed by the colours of the spectrum' and by its frenzied search for the pot of gold at the end of (and at every point along) the spectrum rainbow, the question needs to be asked again about whether citizens will benefit from this pot of gold? Will we be able to dip into it and get some gold too or will it all be snaffled up by the blue sky gold miners? How is the spectrum being divided up? Who will benefit? There are two main players, and bidders: commercial broadcasters and datacasters; both private interests. What of the public interest? Government has taken the role of auctioneer. What is the proper role of government? Should it be auctioneer? Regulator? In the division of spoils of the rainbow-coloured spectrum there seems to be little provision for public-service spectrum.

Writing over 20 years ago, Smythe (1981, pp. 301) argued that 'control of the flow of information [along the channels of communication] is the basis of political power', and economic power, as Smythe (p. 311) later acknowledged: 'the capitalist politico-economic system ... rests upon electronic communication' with 'the radio spectrum as the basis of ... economic, and political power'. Rather than the earth as land, it is the earth as ecosphere with its electromagnetospheric spectrum that is the environmental foundation for the economic base of contemporary capitalism. Writing well before the development of ecommerce, his prescient remarks about those Transnational Corporations 'engaged in the production, sale, leasing and operation of computers' can now be extended to all TNCs today who depend on spectrum in their 'worldwide network of transnational data teleprocessing and storage'.

In the wake of the recent collapse of spectrum auctions in Australia, Stewart Fist (2001, p. 45) has traced the major shift in spectrum management over the last couple of decades:

From licensing to outright sale. This is a form of privatisation. The economic rationalists persuaded governments to treat radio bandwidth as property, so spectrum can now be bought and sold like land, speculated in, and warehoused to create scarcity. Previously, most of us thought of radio bandwidth as part of the common wealth and it was licensed to whoever could use it most productively – socially, as well as economically.

In this context it is productive to consider the US experience. In January 1997 the United States Federal Communications Commission (FCC) opened up the 300 MHz span of the spectrum for telecommunications use and established three bands of spectrum for the Unlicensed-National Information Infrastructure (U-NII) (Jassem, 1998, p. 16). In its decision the FCC stated that it did so as 'to make available spectrum for broadband high data rate unlicensed devices capable of meeting the communication requirements of new multimedia applications' (quoted in Jassem, 1998, p. 16). The use of these low frequency signals 'will largely be limited to the inside of buildings' and could 'permit schools to form inexpensive multimedia networks' (Jassem, 1998, p. 16). Presumably these local radio networks provide cheaper and easier access and are cheaper and easier to maintain than cable networks.

The implications of the FCC decision are profound. Yochai Benkler (1998, p. 294; my emphasis) has argued in an article subtitled 'Building the Commons of the Digitally Networked Environment' that the U-NII band 'raises the possibility that unlicensed wireless devices will provide a component of the information infrastructure that *is not owned by anyone*'. The FCC decision merely enshrines in part the principle that spectrum is a commons that should not be owned by anyone and that all should derive benefit from it equitably. The commons owned by no one is a far cry from the public property of spectrum 'which we all "own"' (Brennan, 1998, p. 791) and the 'open-access' to spectrum espoused by neo-liberal lawyer economists (see Noam, 1998). The commons of the digitally networked environment would be built on the commons of the spectrum. Building the commons of the digitally networked environment would be predicated on the recognition of the commons of the spectrum.

There is a long history of debate in communications' policy circles over the spectrum as commons (see, for example, Brennan, 1998, pp. 791–803). Over 20 years ago Smythe (1981, pp. 309–310, pp. 301–2; see also p. 308) argued that as 'radio frequency assignments are common property..., by international law title to the radio spectrum rests not with individuals or nations, but in all humanity'. Smythe concluded that this places 'the radio spectrum partly in the category of common property on a world scale, and partly in that of state property. Like no other resource, the radio spectrum is the first form of world property.' Benefits from the auctioning or leasing of licences to spectrum should accrue to the whole world, not individual nations. Indeed, the International Telecommunication Union's former Secretary General Richard Butler has argued that 'the 1967 Outer Space Treaty excluded a country from appropriating the profits from space frequencies for itself' (cited in Noam, 1998, p. 775).

The appropriate role of government then would not be to simply sell off bits of the spectrum as a revenue-raising enterprise in order to retire debt balance the budget, or even to make a surplus. Nor would it be to regulate and police every allocation and use of every scrap of spectrum. Rather a more appropriate and useful role would be, as Jassem argues, to be the zoner of the spectrum. In what could be called a 'Declaration of Spectrum Independence' Jassem (1998, p. 22) proposes that governments should

> Declare that the spectrum is large enough to accommodate all users, but complex enough that it will be zoned in order to maximize the benefits accruing from it. Open parts of the spectrum to all users in an unlicensed U-NII-type format. Reserve some spectrum for high powered broadcasters or telecommunication users who will need a license and will be protected from competition from unlicensed users, but who will be required to abide by strong 'public interest' regulations. Use some spectrum for licensed short-term exclusive user, with no pubic interest responsibilities, and with no possibility of renewal. Assign a level of property rights.

Jassem goes on to cite legal precedents in zoning speech and space that could be used for zoning spectrum in the public interest, and for private uses where appropriate. The spectrum is a global commons that has not been, and should not be, enclosed in private property. The mistake made on earth to enclose the commons in private property should not be repeated in the heavens. Jassem (p. 24) reminds us that 'current broadcasters have no property rights to the spectrum they use'. They have licensing rights over leases of the common spectrum owned by all that should be shared equitably by all.

Space power and information superiority

Space is again on the public policy and political agendas with the George W. Bush's resurrection of Ronald Reagan's Strategic Defense Initiative or 'Star Wars' missile shield, albeit under a new name. Yet this move comes as no surprise to observers of US military policy as over the past decade senior military personnel and commentators in the United States have been calling for it to cross 'the final frontier', take 'the ultimate high ground' and become 'masters of space'. Integral to this drive and vital to its success is the deployment and use of communication technologies and the control of flows of information. Both of these take place in orbital, extraterrestrial space, a new front for warfare (as we saw in a previous chapter) and a new medium for the new media of cyberspace and the Internet. In this section, I trace these recent developments and give a critical account of the nationalist and militarist rhetoric in which it is couched. I argue that 'weaponisation' of space is in contravention of a number of international treaties. I conclude that 'astroenvironment-alism' should be a broadly based popular movement of resistance to these moves and of action for the global commons of space owned by none and shared by all.

George W. Bush's announcements about National Missile Defence (NMD) – a new declaration of Star Wars – brings home that extrater-restrial space is being colonised by nation-states (or by one nation-state in particular) and its commons enclosed in the private property of corporations. In particular, the privatisation and militarisation (or more precisely 'weaponisation') of extraterrestrial space proceeds apace despite the Outer Space Treaty of 1967 adopted by a majority of United Nations and re-affirmed in November 2000, which seeks to 'set aside' extraterrestrial space as a kind of inter-national park for all nations to share in and to be owned by none. The 1979 Moon Treaty argued along similar lines for 'commonality of ownership of space bodies' (Marshall, 1999). Such moves come in response to US attempts to become 'Masters of Space'.

After the militarisation of space in the 1980s comes its 'weaponisation' in the 2000s. In the meantime and in the aftermath of the Gulf War in 1991 there was a growing recognition that 'future US security hinges on [a] dominant role in space' (Berkowitz, 1992, p. 71). If it does not maintain that role (and even if it does), 'a space-based Pearl Harbour could be around the corner' (Berry, 2001). Ironically, in the same week in which the jingoistic film *Pearl Harbour* was released *New Scientist* published an article by Oberg (2001, p. 28) in which he claimed that 'without more funding for space defence, the US faces a Pearl Harbour in orbit'. Perhaps the irony goes beyond mere coincidence into the realms of conspiracy, though there is probably nothing as sinister and conjectural operating here. Rather, what could be called 'condiscursivity' is operating in which a common rhetoric of paranoia and xenophobia is deployed across the cultural landscape for reactionary political ends. Exactly who or what would be attacked, and who or what would do the attacking is not specified in either case, but raising this spectre from the past may be enough to prompt national action on this front, or perhaps more precisely frontier.

Fascination with the frontier – crossing it, closing it and cloning it – is a defining feature of American culture. The frontier is either terrestrial, as it was with the winning of the national territory from the indigenes during the seventeenth to nineteenth centuries, or extraterrestrial as with the conquest of space in the 1960s. It can even be invoked to figure science as Vannevar Bush (1945) did in the dying days of the Second World War in a report to President F. Roosevelt commissioned by him. 'The endless frontier' of science, as Bush put it, presented to post-war America the opportunity to continue the epistemological and institutional colonisation of nature, to extend the boundaries of the empire of nature into hitherto unexplored wildernesses and to develop knowledge of and exercise power over it.

The famous historian Frederick Jackson Turner argued in 1893 that (quoted by Healy, 1997, pp. 55 and 6; see also Turner, 1893/1961) 'American social development has been continually beginning over again on the frontier. This perennial rebirth, this fluidity of American life ... furnish the forces dominating the American character.' Turner was weighing the significance of the census of 1890, which for him indicated that the frontier was closed (Kern, 1983, p. 164). Interestingly, and ironically, the 1890 census also marked another turning point: the first large-scale use of mechanical computation for social calculation with the use of Hollerith's punch-card machine, a precursor of the computer (Shurkin, 1984, p. 78; Hanson, 1982, p. 48; Campbell-Kelly and Aspray, 1996, pp. 20–26).

The fully-fledged computer later converged with telecommunications to produce cyberspace with its own frontier mythology of the 'electronic frontier', the 'information frontier' and the 'Internet frontier' even with its 'frantic Wild West atmosphere' and pioneers (Gleick, 2002, subtitle and pp. 45, 70, 90 and 260; see also Healy, 1997). The computer and telecommunications also opened the extraterrestrial frontier of orbital space. Despite, or perhaps because of, a flood of jingoistic nationalism surrounding it, the frontier mythology has not been without its critics. In the 1950s poet and doctor William Carlos William saw American history as 'full of a rich regenerative violence'. In a trilogy on the mythology of the frontier whose first volume quoted Williams as an epigraph and invoked in its title his idea of 'regeneration through violence', Slotkin (1973; 1992; 1994) critiqued Turner's masculinist 'frontier thesis', especially the typology of the frontier hero.

Slotkin's great trilogy ends in 1973 with the fall of Saigon and the defeat of the United States in the slimy swamps of Vietnam (see Giblett, 1996, chapter 9). Yet this ignominious defeat by no means heralded the end of the mythology of the frontier but the beginning of its displacement into sublime extraterrestrial space and cyberspace (and victory in the 1991 Gulf War18 years later) and the perpetuation of the mythology of the frontier hero in the astronaut, console cowboy, hacker, videogamer and web-surfer. Regeneration through violence was sublimated into regeneration through video violence. Healy (1997, p. 57) sees the Internet as 'a new kind of frontier' with the pioneer in cyberspace as the new frontier hero opening it up for what Fisher (1997, p. 121) calls 'mass migration into cyberspace'. Cyberspace is a new frontier replacing the old new frontier of orbital extraterrestrial space and pointing towards the future frontier of extra-orbital space ('extrorbitant' space).

Extra-orbital space ('ex(tr)orbitant' space) is the future frontier for the United States intent on pursuing its frontier mythology into new territories (Oberg, nd, pp. 166–168). 'Let us conquer space' was the rallying cry in the nineteenth century for crossing the terrestrial frontier and entering the 'howling wilderness' beyond (Oberg, nd, p. 1). National frontier mythology was embodied in the national heroes of the frontier such as Daniel Boone and Davy Crockett as it was later to be in John Glenn and Neil Armstrong. The frontier purportedly welded the nation together and created a sense of national purpose. The closing of the territorial frontier was a defining moment in national history just as the opening of the extraterrestrial frontier was a new moment of national definition and inaugurated a new sense of national destiny.

Yet with the closing of the orbital frontier a new extra-orbital frontier was sought beyond it. The Mars Society advocates human settlement on that planet in terms of the frontier mythology by attributing 'current social ills to be a consequence of the loss of the frontier' (Oberg, nd,

p. 167). Mars is the new frontier whose opening, and 'homesteading' as Lewis (1996, pp. 155–172) puts it in similar vein, will miraculously cure all social ills on earth. As Marshall (1999) argues, 'frontiersmen never die, they just drift off into space', for them 'the final frontier'. He goes on to list eight titles of recent articles and books devoted to the 'space frontier' and 'space colonies'.

Just as US Congressmen in the nineteenth century proposed the paving of a road through the Cumberland Gap, likewise for (Oberg, nd, p. 1) 'federal investments in and subsidies of canal, railroad, ... aircraft and so forth have opened doors and lowered thresholds for public and commercial traffic to flow through'. Oberg and others see a similar scenario being played out in extra-orbital space now and into the foreseeable future. The US military, according to Grossman (2001, p. 34), wants to 'control space' and 'dominate' the earth as 'explicitly stated in documents of the US Space Command' (USSPACECOM). Its *Vision for 2020* (online) proclaims that 'today, the United States is the pre-eminent military space power'. The reason? 'USSPACECOM is the only military organization with operational forces in space.'

Its website http://www.spacecom.af.mil/usspace/ explains that it was set up in 1985 to 'help institutionalise the use of space ..., the ultimate high ground'. Objects in space have, as Oberg (nd, p. 4; see also p. 14) puts it, 'a VANTAGE point for viewing large areas on the ground'. From this vantage point and high military ground the US military wants to wage future wars on earth. On this military high ground it may also wage future wars in space (General Joseph Ashy cited by Grossman, 2001, p. 37), which some see as 'virtually inevitable' (see Oberg, nd, p. 61). Rather than being a matter of *if* warfare moves to space, for Oberg (nd, p. 129), it is a only a matter of '*when* warfare moves to space'.

The threat of space war is the post–Cold War, as it, like its predecessor with nuclear war, verges on inevitability. Yet, unlike the balance of terror that characterised the Cold War between the two antagonists of United States and the USSR, the post–Cold War characterised by what Virilio (2000a, p. 19; 2000b, pp. 5, 58) calls 'an imbalance or unbalance of terror' is being fought between the United States and whoever fits the bill of a 'rogue state' or harbourer of terrorists. Also unlike the balance of terror between the two nuclear superpowers, the post–Cold War is being fought between the United States with the instruments of terror (communication satellites) above our heads in orbital extraterrestrial space and terrorists without the same space arsenal who resort to other means as we have seen recently in using civilian aircraft as bombs and civilian mail for chemical warfare.

As war in space is 'virtually inevitable' so also (Oberg, nd, pp. 61 and 129) 'the weaponisation of space is inevitable'. Just as the Cold War deployed nuclear weapons, so 'Space War' deploys space weapons. Under the Outer Space Treaty of 1967, as Oberg (nd, p. 81) puts it, 'warlike activities are forbidden in space and on celestial bodies, save in self-defence or defence of allies'. Military surveillance, information processing and ground targeting from space could be construed as warlike activities, or they could be condoned as defensive ones. Either way, it is a moot point as the nexus between communication technologies and the military is well established.

The association between communication technologies (such as telegraphy, photography and cinematography) and war is a long and necessary one. Whilst some communication technologies were developed for military purposes (such as satellites), the development of other communication technologies (such as television) was hindered because there was no immediate military use, or investment as we have seen previously. Recent developments of Space and Star Wars are no exception to this nexus. Communication technologies, just as much as guns, are weapons of war as we have also seen previously.

Extraterrestrial space may have no end, but where does it begin? Oberg (nd, pp. 79–80) points out that

> even after decades of space activities, there is still no legal definition of where 'space' begins and national sovereignty ends. Although maritime boundaries tended to originally be defined by the range of naval gunfire, the ability of several nations to attack low-orbit objects has not led to an extension of national sovereignty to these altitudes.

Extraterrestrial space may begin as low as about 30 kilometres above the earth or as high as about 160 kilometres, or even further beyond 35,800 kilometres. The Bogota Declaration of 1976 signed by 8 equatorial countries declared that 'geostationary orbit [at about 35,800 kilometres above the equator] is a *scarce natural asset* that is not part of outer space' and that 'the geostationary orbit arc above each country is the sovereign territory of the country' (Oberg, nd, pp. 72 and 100; my emphasis). It also declared that 'the geostationary arc above the oceans are part of the common heritage of all mankind and should be exploited to the benefit of all mankind'.

The United States has other plans. The *Long Range Plan* (online) of the US Space Command defines 'control of space' as the ability to secure access to space, freedom of operations within the space medium, and an ability to deny others the use of space, if required. Achieving and maintaining Control of Space will influence all national and military objectives. Future space programs will be "consumer oriented" to assure information dominance of the warfighter. This operational concept encompasses today's missions of space control and space support (http://www.peterson.af.mil/usspace/LRP/; see also Oberg, nd, p. 10)

Extraterrestrial space is figured as a 'medium', rather than being seen as a place, let alone as a commons. For Oberg (nd, p. 4; see also pp. 126–127), 'space used to be a barrier, but like the oceans, it is being transformed into a medium for transportation and a medium for harvesting'. Just as communication followed transportation as the leading-edge technology and industry, and communication became separated from transportation (as we have seen previously), so space becomes a communication medium for harvesting benefits, not only in spectrum auctions but also in the deployment of satellites and space weaponry. Space, like the oceans, is not being seen as an ecosystem, as a habitat, as part of the ecosphere. Rather it is being seen as a field in and from which the United States can, as Oberg (nd, p. 86) puts it, 'reap the benefits of space power'. Space, like the oceans, has been militarised and is being weaponised.

Extraterrestrial space is seen as a medium because it is not just spatial but also temporal. Oberg (nd, p. 126) argues that 'access to the medium of space has already changed the conventional terrestrial concepts of area, volume and time'. It has made possible extensive surveillance of vast tracts of the earth's surface (Oberg, nd, p. 124) and of large volumes of air space. It has also made possible almost instantaneous communication in 'real time' across large distances with the twin result of what Brown (1996, p. 33) sees as time-compression and space-distortion.

Space-based communications come at the end of a long history of time-compression and space-distortion. Dearth and Goodden (1996, p. 274) trace this history:

Since the early days of the Industrial Age, time increasingly has been compressed; yet it has become more important... Initially, it was transportation technology that drove this compression. The steam engine was applied to ground and then marine transport. Later the telegraph – and then wireless telegraphy – enabled *information* to be passed over greater distances with amazing speed. There followed the telephone and its adjunct – actually a technology that pre-dated the telephone – the facsimile machine. Then it was television, which introduced the added dimension of more vivid graphical – and life-like – images. More recently, it has been the computer – and particularly the microprocessor – linked via local – and wide – area networks. Thus, speed has increased ever more dramatically as a function of distance. The result is that *distance means less – and, hence, time is compressed* (their emphasis).

Extraterrestrial space is also a medium because it is where the threat of war is mediated. The medium of space is what Oberg (nd, p. 147 my emphasis) calls 'an emerging linchpin for the *threat* and application of force and of the conduct of war'. Space is no longer the sky above our heads where the peaceful spirits of the earth fly, nor the heavens where the gods reside, but the medium through which the threat of war in space is mediated.

Just as the United States is the world's greatest seafaring nation it is also 'the world's greatest spacefaring nation' according to General Howell M. Estes III, former Commander in Chief of the US Air Force and US Space Command, writing in his foreword to *Space power theory* (Oberg, nd, p. x). For Oberg (nd, p. 49), 'the United States holds a dominant lead in Earth's space activities' as it has 'the most far-flung fleet of interplanetary space probes in the history of the space age'. It is also the world's greatest military space power.

China is challenging this supremacy, or at least the United States perceives a challenge to its supremacy coming from that quarter. The Commission to Assess US National Security Space Management and Organization is quoted as arguing that 'the Xinhua news agency reported that China's military is developing methods and strategies for defeating the US military in a high-tech and space-based future war' (Berry, 2001). The threat may be more perceived than real with space just another theatre of operations. Just as the oceans became the stage on which nations could flex their military muscle and boast of their status as powers, so space is becoming a similar domain. According to Oberg (nd, p. 63), 'the Chinese government has obviously selected space operations as an area to prove their status as a modern great power'. Space is reduced to merely the context in which to prove great power status and the pretext for space operations. Space is seen having no inherent worth or other uses.

Despite the militarisation of space and the military beginnings of spaceflight, the boosters of US space conquest laud the 'peaceful' uses of space in biblical terms. Oberg (nd, p. 6) acknowledges that 'spaceflight sprang from MILITARY roots'. He later concedes that space travel had a 'far-from-immaculate conception' (p. 143). Yet it did have an immaculate conception in the Bachelor Machine for a Bachelor Birth of the military–industrial–university complex. Bachelor Machines circumvent not only the body of the mother, but also the earth. Spaceflight enacts the desire to escape from 'the prison of the earth' and 'orbit in the sublime company of heavenly bodies' (as we have seen previously). Spaceflight had the same immaculate conception as the nuclear bomb (see Easlea, 1983) and as a long line of Bachelor, or celibate, machines immaculately conceived before it, including communication technologies such as photography, cinema, radio, television, satellites and the computer (as we have also seen previously).

Despite these military roots and conception, spaceflight, for Oberg (nd, p. 6), is

> now surprisingly 'peaceful' – possibly the most genuine 'swords into ploughshares' metamorphosis in history. However, those plough-shares can also quickly change back into swords. The 1991 Gulf War demonstrated the exceptional military utility of space systems.

Oberg (nd, p. 123) goes on later to argue that for some within the US military 'the [1991] Persian Gulf War represents the first space war' (see Anson and Cummings, 1992, pp. 121–133). Yet he finds this claim 'dubious' as 'there have been no warriors in space; there have been no weapons fired from space against terrestrial targets; and, there have been no space-to-space engagements' (p. 121). But there have been and are military surveillance and targeting satellites in space; there have been weapons fired against terrestrial targets targeted from space; and so there have been space-to-earth engagements. Although 'space remains unarmed' (p. 125), it is not unmilitarised. The 1991 Gulf War may not have been 'the first space war' but it was, as (Anson and Cummings (1992, p. 121) put it, 'the first occasion in which a full range of military space systems was used in anger', a euphemism for war.

Just as a tree can never sever its connection from its roots and survive, so spaceflight can never get away from its military beginnings. Swords can never be beaten into ploughshares without bearing the traces of their former occupation. If a ploughshare can be turned back into a sword, it has never entirely ceased to be a sword. The technology is the use, the first use. The military roots of the ploughshare produce the benefits that can be reaped in harvesting space. The ploughshare 'breaks the plains' of space just as it did the earth in the agricultural conquest of the 'wilderness' of the prairie. Celesti-culture succeeds agriculture.

Just as telecommunications, computers and satellites have converged so do the soldier, sailor, marine and pilot converge in the 'warfighter' Whereas the soldier, sailor, marine and pilot were partially defined by the front or the terrain on which, and terrestrial space in which, they fought, the warfighter fights in extraterrestrial space where all wars against the United States are now lost (or at least achieve a state of unaccomplishment as (and beginning) with the 1991 Gulf War as we have seen) and is defined purely by the function he (or she) serves.

The warfighter is also defined by 'the medium' of space in which he/she serves rather than by the front on which he/she serves as 'the medium of space is the fourth medium of warfare – along with [and after] land, sea, and air' as the US Space Command *Vision for 2020* puts it. As I have argued previously, space is the fifth front of warfare after land, sea, air and spectrum. Just as 'air support' evolved into 'air power' to protect 'US national interests and investments' so will space support 'evolve' into space power to do the same. The social Darwinism of the 'survival of the fittest' is extrapolated from the 'jungles' of the industrial city and projected onto extraterrestrial space just as Darwinian evolution did onto equatorial rainforests (see Giblett, 2004, chapter 12). In *Vision for 2020* (online) General Estes has the 'US Space Command dominating the space dimension of military operations to protect US interests and investments' both on earth and in space, or in the heavens. 'The American spatial high command', as Virilio (2002a, p. 2) dubs it, is 'a true *deus ex machina* of planetary peace or war'. The US Space Command is a capricious Greek god of Greek tragedy who rules earth from the Mt Olympus of extraterrestrial space and throws thunderbolts at lightning speed to interfere in the affairs of earthlings.

Vision for 2020 predicts that 'during the early portion of the twenty-first century, space power will...evolve into a separate and equal medium of warfare'. The United States may even 'evolve into the guardian of space commerce – similar to the historical example of navies protecting sea commerce' as 'space commerce is becoming increasingly important to the global economy'. Gunboat 'diplomacy' will then extend into space with spaceship diplomacy in which the United States will play guardian to the colonies of space, to its colonies in space, and on earth.

The United States also wants to act as a steward in managing the estates of space by reaping its harvests and distributing its rewards. The US military wants to take not only the high spatial ground, but also the high moral ground, and to couch and justify its claims to the former in terms of the latter. General Estes III proclaims that 'as stewards for military space, we must be prepared to exploit the advantages of the space medium' just as the good Christian stewards of terrestrial space exploited the advantages of the earth 'medium' and 'harvested' its goods.

Information is crucial for doing so. Information is crucial for controlling space and controlling space is crucial for 'information superiority'. Communication satellites and other communication technologies such as radio are the conduits and vectors for the control and flow of information. The electromagnetic spectrum is the 'natural resource' exploited for doing so. The launching and positioning of communication satellites in extraterrestrial space are vital for communicating information and for maintaining pre-eminence over terrestrial sites. An information circuit is set up between the terrestrial target, extraterrestrial communication satellites and weapons. The flow of information from the target via the satellite to the weapon is transformed into a counter-flow of deadly force against the target aimed by the satellite. Controlling the military high ground of extraterrestrial space becomes crucial for controlling the military low ground of terrestrial space.

'Control of Space' is seen as 'essential to achieving the force multiplying effect of Information Superiority' defined in both the *Long Range Plan* and *Vision 2020* as 'the capability to collect, process, and disseminate an uninterrupted flow of information while exploiting or denying an adversary's ability to do [or 'leverage' as *Vision* puts it] the same'. The *Long Range Plan* goes on to argue that 'space superiority is essential to Information Superiority...Threats to space are threats to Information Superiority.' What Zuboff (1988, p. 322) calls 'the panoptic power of information technology' is no longer confined to terrestrial space but is increasingly extended to extraterrestrial space. 'Surveillance', 'intelligence' and 'reconnaissance' (ISR) become key terms superseding the C3 ('command', 'control' and 'communication') of previous wars as the military mantra of mastery (see Brown, 1996, p. 31).

Rather than attacking the enemy's external defences first, and as the 1991 Gulf War amply illustrated, 'strategic information warfare' involves attacking, as Brown (1996, p. 39) puts it, 'the more vulnerable internal systems first'. These include the power grid and telecommunications with the aim of preventing, in the words of a US Air Force Colonel (cited in Brown, 1996, p. 40), 'the system's leadership from gathering, processing, and using information we don't want him to have'. Control of space means not only gaining the spatial high ground, but also, as Brown (1996, p. 43) puts it, 'battling for the information high ground'.

The space warrior is also an 'information warrior' (Brown, 1996, p. 46). In fact, 'space warriors defend information assets' (Sheehy, 2001).

There are three types of Information Warfare according to Brown (1996, p. 47):

> Type I Information Warfare involves managing the enemy's perceptions through deception operations, psychological operations, what the Joint Staff calls 'Truth Projection,' and a variety of other techniques. At the same time, one must protect against enemy perception management efforts ... Type II Information Warfare involves denying, destroying, degrading or distorting the enemy's information flows in order to break down his organizations and his ability to coordinate operations [such as destroying the Iraqi power grid and telecommunications as in the 1991 Gulf War] ... Type III Information Warfare gathers intelligence by exploiting the enemy's use of information systems. A much larger challenge, however, may be in protecting friendly information systems from exploitation by other intelligence organizations.

With this last type, the Information Warrior would tap into, and eavesdrop on, the enemy's military communications in order to glean secrets whilst at the same time trying to stop the same courtesy being exercised against him and his allies. The Information Warrior would be involved in both offence and defence and would therefore be a fully-fledged fighter, and information warfare would be (a part of) war. Brown (1996, p. 49) concludes direly that 'a nation without an Information Warfare capability will be a nation without a military capability'.

Space power, for Oberg (nd, p. 127), must be 'combined' with its 'emerging sibling' of information power. Whereas communication technologies were merely the stepchildren of war, Oberg sees information power as the younger sibling of space power. The younger sibling, for him, should combine its forces with the older. Space power depends on space surveillance. For Oberg (nd, p. 14), 'surveillance of space emerges as the key element of space control, enabling the other facets of protection and denial'. The *Long Range Plan* stresses that 'to assure access to space, we must surveil it. *Surveillance of space* allows total battlespace awareness, freedom of operations, and deconfliction of activities to, in, and from space – the cornerstones to "enforcing the peace".' Figure 11.1 shows somewhere in northern South America marked as a 'no-go' area (indicated thus '⊘') and being zapped by a red thunderbolt from on high in a secularised replay of the power of god.

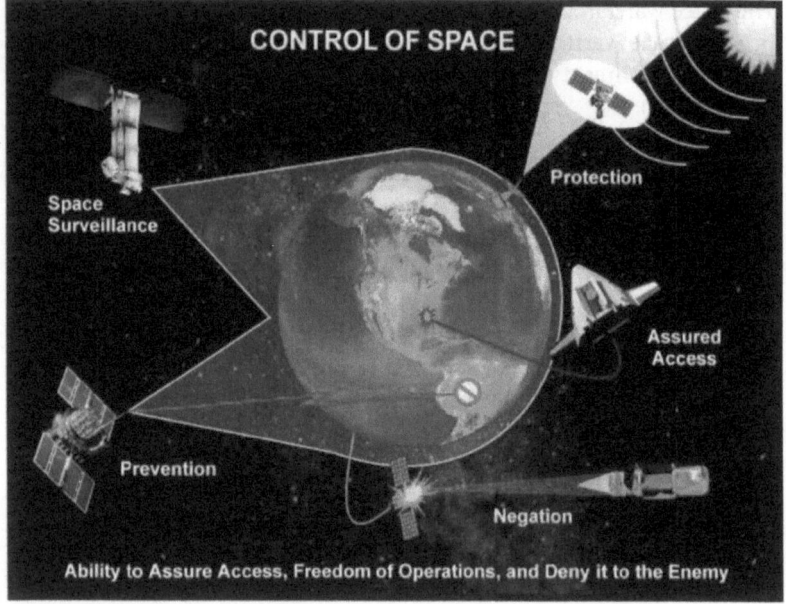

Figure 11.1 Concepts of operations for control of space (according to US Space Command)

In the chapter entitled 'The Vision: Focused on the Warfighter' the *Plan* maintains that

> If we don't have unfettered access to our space capabilities, we can't secure the 'high ground.' If we don't secure the high ground, we assume substantially greater risk when we try to successfully manoeuvre, strike, or adequately protect and sustain our forces. Control of Space ensures our forces have situational understanding and denies that product to the enemy. Our advantage in seeing and understanding the evolving situation, coupled with the enemy's inability to see, enables Dominant Manoeuvre, Precision Strike, Full-Dimensional Protection and Focused Logistics.

Extraterrestrial space is seen as just another high ground to take, another post from which to command, another frontier to cross, another place to colonise, another commons to enclose, another eminence for pleasing prospects, another resource to exploit, another commodity to sell – all while preventing competitors, national and commercial, from doing so.

'The heart of future space operations', for Oberg (nd, p. 161), 'has to be the industrialization of space', including mining operations on the moon and the asteroids 'which may lead to the colonization of space' in the sense of human settlements. Extraterrestrial space is becoming the high commercial ground with the development of 'capitalism in space' (Grossman, 2001, p. 37) in 'the age of space commercialisation'.

Resisting these moves is 'astroenvironmentalism', a term coined by Ryder Miller (cited in Grossman, 2001, p. 37) to describe the calls for a bureaucratic environmentalism of 'impact statements' in extraterrestrial space. Yet astroenvironmentalism should also be a popular environmentalism of common ownership in extraterrestrial space. Rather than the commons of extraterrestrial space being enclosed in the public property of nations and private property of corporations, all should have equal ownership of, and access to, its goods and benefits. All should take equal responsibility for its care. After all, our earthly home is not just terrestrial but extraterrestrial in the electromagnetosphere and orbital space.

References

Adorno, T. and W. Benjamin (1999). *The Complete Correspondence: 1928–1940*. H. Lonitz (Ed.) (N. Walker, Trans.). Cambridge, Massachusetts: Harvard University Press.

Aitken, H. (1976). *Syntony and Spark: The Origins of Radio*. New York: John Wiley & Sons.

Anson, P. and D. Cummings (1992). The first space war: The contribution of satellites to the Gulf War. In A. Campen (Ed.), *The First Information War: The Story of Communications, Computers and Intelligence in the Persian Gulf War* (pp. 121–133). Fairfax: AFCEA International Press.

Arendt, H. (1958). *The Human Condition*. Chicago: University of Chicago Press.

Aristotle (1984). *The Complete Works: Volume 1, The Revised Oxford Translation*. J. Barnes (Ed.). Princeton: Princeton University Press.

Bailes, H. (1980). Military aspects of the war. In P. Warwick (Ed.), *The South African War: The Anglo-Boer War 1899–1902* (pp. 65–102). London: Longman.

Ballard, J. G. (1973/1995). *Crash*. London: Vintage.

Balsamo, A. (1995). Forms of technological embodiment: Reading the body in contemporary culture. In M. Featherstone and R. Burrows (Eds), *Cyberspace/Cyberbodies/Cyberpunk: Cultures of Technological Embodiment* (pp. 215–237). London: Sage.

Barbusse, H. (1916/2004). *Under Fire* (R. Buss, Trans.). London: Penguin.

Barthes, R. (1972). The world as object. *Critical Essays* (pp. 3–12) (R. Howard, Trans.). Evanston: Northwestern University Press.

Barthes, R. (1973). *Mythologies* (A. Lavers, Trans.). London: Granada.

Barthes, R. (1979). Power and 'cool'. *The Eiffel Tower and Other Mythologies* (pp. 43–45) (R. Howard, Trans.). New York: Hill & Wang.

Barthes, R. (1981). *Camera Lucida: Reflections on Photography* (R. Howard, Trans.). New York: Farrar, Straus, Giroux.

Barty-King, H. (1979). *Girdle Round the Earth: The Story of Cable and Wireless and its Predecessors*. London: Heinemann.

Batchen, G. (1997). *Burning with Desire: The Conception of Photography*. Cambridge, Massachusetts: The MIT Press.

Baudrillard, J. (1975). *The Mirror of Production* (M. Poster, Trans.). St Louis: Telos.

Baudrillard, J. (1981). *For a Critique of the Political Economy of Sign* (C. Levin, Trans.). St Louis: Telos.

Beer, G. (1996). 'Wireless': Popular physics, radio and modernism. In J. Uglow and F. Spufford (Eds), *Cultural Babbage: Technology, Time and Invention* (pp. 149–166). London: Faber & Faber.

Bellamy, D. (1998). Seminar, Centre for Ecosystem Management, Edith Cowan University, Joondalup Campus.

Beniger, J. (1986). *The Control Revolution: Technological and Economic Origins of the Information Society*. Cambridge, Massachusetts: Harvard University Press.

Benjamin, W. (1973a). *Charles Baudelaire: A Lyric Poet in the Era of High Capitalism* (H. Zohn, Trans.). London:Verso.

Benjamin, W. (1973b). *Illumination* (H. Zohn, Trans.). London: Fontana.

Benjamin, W. (1979). *One-way Street and Other Writings* (E. Jephcott and K. Shorter, Trans.). London: NLB.

Benjamin, W. (1986). *Moscow Diary*. G. Smith (Ed.) (R. Sieburth, Trans.). Cambridge, Massachusetts: Harvard University Press.

Benjamin, W. (1994). *The Correspondence of Walter Benjamin: 1910–1940*. G. Scholem and T. Adorno (Eds) (M. Jacobson and E. Jacobson, Trans.). Chicago: University of Chicago Press.

Benjamin, W. (1996). *Selected Writings: Volume 1, 1913–1926*. M. Bullock and M. Jennings (Eds). Cambridge, Massachusetts: The Belknap Press of Harvard University Press.

Benjamin, W. (1999a). *The Arcades Project* (H. Eiland and K. McLaughlin, Trans.). Cambridge, Massachusetts: The Belknap Press of Harvard University Press.

Benjamin, W. (1999b). *Selected Writings: Volume 2, 1927–1934*. M. W. Jennings, H. Eiland and G. Smith (Eds) (R. Livingston and others, Trans.). Cambridge, Massachusetts: The Belknap Press of Harvard University Press.

Benjamin, W. (2002). *Selected Writings: Volume 3, 1935–1938*. H. Eiland and M. Jennings (Eds) (E. Jephcott, H. Eiland and others, Trans.). Cambridge, Massachusetts: The Belknap Press of Harvard University Press.

Benjamin, W. (2003). *Selected Writings: Volume 4, 1938–1940*. H. Eiland and M. Jennings (Eds) (E. Jephcott and others, Trans.). Cambridge, Massachusetts: The Belknap Press of Harvard University Press.

Benkler, Y. (1998). Overcoming agoraphobia: Building the commons of the digitally networked environment. *Harvard Journal of Law and Technology, 11* (2), 287–400.

Benthall, J. (1976). *The Body Electric: Patterns of Western Industrial Culture*. London: Thames & Hudson.

Bentham, J. (1787/1962). Panopticon; or, the inspection-house. *Works: Volume 4* (pp. 35–170). New York: Russell & Russell.

Berkowitz, M. (1992). Future US security hinges on dominant role in space. *Signal*. May, 71.

Berman, M. (1983). *All that is Solid Melts into Air: The Experience of Modernity*. London: Verso.

Berry, S. (2001). Reports examine US space-faring survival. *Signal*. http://www.us.net/signal/CurrentIssue/April01/space-april.html

Bey, H. (1996). The information war. In T. Druckrey (Ed.), *Electronic Culture: Technology and Visual Representation* (pp. 369–375). New York: Aperture.

Blunden, E. (1928). *Undertones of War*. London: Penguin.

Bolen, J. (1984). *Goddesses in Everywoman: A New Psychology of Women*. San Francisco: HarperCollins.

Boyle, R. (1675/1966). Experiments and notes about the mechanical origin or production of electricity. In T. Brown (Ed.), *Works: Volume 4* (pp. 345–350). Hidesheim: Georg Olms Verlagsbuchhandlung.

Brennan, T. (1998). The spectrum as commons: Tomorrow's visions, not today's prescription. *Journal of Law and Economics, XLI* , 791–803.

Brown, D. (1970). *Bury My Heart at Wounded Knee: An Indian History of the American West*. London: Barrie & Jenkins.

Brown, M. (1996). The revolution in military affairs: The information dimension. In A. Campen, D. Dearth and R. Gooden (Eds), *Cyberwar: Security, Strategy and Conflict in the Information Age* (pp. 31–52). Fairfax, VA: AFCEA International Press.

Brown, N. (1959). *Life Against Death: The Psychoanalytical Meaning of History.* Hanover, NH: Wesleyan University Press.

Buber, M. (1970). *I and Thou* (3rd ed.) (W. Kaufmann, Trans.). Edinburgh: T & T Clark.

Bullfinch, T. (1993). *The Golden Age of Myth and Legend.* Ware: Wordsworth.

Burke, D. (1991). *Road Through the Wilderness: The Story of the Transcontinental Railway: The First Great Work of Australia's Federation.* Kensington, NSW: University of New South Wales Press.

Bush, V. (1945). *Science: The Endless Frontier: A Report to the President.* Washington: United State Government Printing Office.

Cadava, E. (1992). Words of light: Theses on the photography of light. *Diacritics, 22*(3–4), 84–114.

Caldwell, L. K. (1984). *International Environmental Policy: Emergence and Dimensions.* Durham: Duke University Press.

Campbell-Kelly, M. and W. Aspray (1996). *Computer: A History of the Information Machine.* New York: Basic Books.

Campen, A. (Ed.) (1992a). Introduction. *The First Information War: The Story of Communications, Computers and Intelligence in the Persian Gulf War* (pp. ix–xxi). Fairfax: AFCEA International Press.

Campen, A. (Ed.) (1992b). Silent space warriors. *The First Information War: The Story of Communications, Computers and Intelligence in the Persian Gulf War* (pp. 135–141). Fairfax: AFCEA International Press.

Carey, J. (1989). Technology and ideology: The case of the telegraph. *Communication as Culture: Essays on Media and Society* (pp. 201–230). Boston: Unwin Hyman.

Carey, J. with Quirk, J. (1989a). The mythos of the electronic revolution. *Communication as Culture: Essays on Media and Society* (pp. 113–141). Boston: Unwin Hyman.

Carey, J. with Quirk J. (1989b). The history of the future. *Communication as Culture: Essays on Media and Society* (pp. 173–200). New York: Routledge.

Carroll, B. (1992). *Australian Communications Through 200 Years.* Kenthurst, NSW: Kangaroo Press.

Carrouges, M. (1954). *Les Machines Célibitaires.* Paris: Arcanes.

Carrouges, M. (1975). Directions for use. *Le Machine Celibi/the Bachelor Machines* (pp. 21–49). New York: Rizzoli.

Carter, P. (1989). *The Road to Botany Bay: An Exploration of Landscape and History.* Chicago: University of Chicago Press.

Caygill, H. (1998). *Walter Benjamin: The Colour of Experience.* London: Routledge.

Certeau, M. de (1983). The madness of vision. *Enclitic, 7*(1), 24–31.

Certeau, M. de (1984). *The Practice of Everyday Life* (S. F. Rendall, Trans.). Berkeley: University of California Press.

Certeau, M. de (1986). The arts of dying: Celibatory machines. *Heterologies: Discourse on the Other* (pp. 156–167) (B. Massumi, Trans.). Manchester: Manchester University Press.

Certeau, M. de (1987). The gaze: Nicholas of Cusa. *Diacritics, 17*(3), 2–38.

Chanan, M. (1980). *The Dream that Kicks: The Prehistory and Early Years of Cinema in Britain*. London: Routledge & Kegan Paul.

Chandler, Jr., A. (1977). *The Visible Hand: The Managerial Revolution in American Business*. Cambridge, Massachusetts: Belknap Press of Harvard University Press.

Christie, I. (1994). *The Last Machine: Early Cinema and the Birth of the Modern World*. London: BBC Educational Developments.

Cole, T. (1835/1965). Essay on American scenery. In J. McCoubrey (Ed.), *American Art: 1700–1960: Sources and Documents* (pp. 98–110). Englewood Cliffs: Prentice–Hall.

Comolli, J-L. (1980). Machines of the visible. In T. de Lauretis and S. Heath (Eds), *The Cinematic Apparatus* (pp. 145–153). London: Macmillan.

Conley, V. A. (1997). *Ecopolitics: The Environment in Poststructuralist Thought*. New York: Routledge.

Cornwell, J. (2003). *Hitler's Scientists: Science, War and the Devil's Pact*. London: Penguin.

Cowan, R. (1997). *A Social History of American Technology*. New York: Oxford University Press.

Creveld, M. van (1985). *Command in War*. Cambridge, Massachusetts: Harvard University Press.

Creveld, M. van (1991). *Technology and War: From 2000 B.C. to the Present* (Revised & expanded ed.). New York: Free Press.

Crowley, D. and P. Heyer (1991). *Communication in History: Technology, Culture, Society*. New York: Longman.

Cubitt, S. (1998). *Digital Aesthetics*. London: Sage.

Cubitt, S. (2000). Virilio and new media. In J. Armitage (Ed.), *Paul Virilio: From Modernism to Hypermodernism and Beyond* (pp. 127–142). London: Sage.

Cunningham, S. (1997). Television. In S. Cunningham and G. Turner (Eds), *The Media in Australia* (2nd ed., pp. 90–111). Sydney: Allen & Unwin.

Czitrom, D. (1982). *Media and the American Mind: From Morse to McLuhan*. Chapel Hill: University of North Carolina Press.

Dadley, P. (1996). The garden of Edison: Invention and the American imagination. In J. Uglow and F. Spufford (Eds), *Cultural Babbage: Technology, Time and Invention* (pp. 81–98). London: Faber & Faber.

Danly, S. (1988). Introduction. In S. Danly and L. Marx (Eds), *The Railroad in American Art: Representations of Technological Change* (pp. 1–50). Cambridge, Massachusetts: The MIT Press.

Dearth, D. and R. Goodden (1996). Epilogue. In D. Dearth, R. Goodden and A. Campen (Eds), *Cyberwar: Security, Strategy and Conflict in the Information Age* (p. 274). Fairfax, VA: AFCEA International Press.

Deleuze, G. and F. Guattari (1977). *Anti-Oedipus: Capitalism and Schizophrenia* (R. Hurley, M. Seem and H. R. Lane, Trans.). New York: Viking.

Demac, D. (1986a). Communication satellites and the third World. In M. Traber (Ed.), *The Myth of the Information Revolution: Social and Ethical Implications of Communication Technology* (pp. 35–45). London: Sage.

Demac, D. (Ed.) (1986b). Introduction. *Tracing New Orbits: Cooperation and Competition in Global Satellite Development* (pp. xi–xvii). New York: Columbia University Press.

Demac, D. (1990). New communication technologies: A plug 'n play world? In J. Downing, A. Mohammadi and A. Sreberny-Mohammadi (Eds), *Questioning the Media: A Critical Introduction* (pp. 207–216). London: Sage.

Dickson, P. (1976). *The Electronic Battlefield*. Bloomington: Indiana University Press.

Douglas, M. (1966). *Purity and Danger: An Analysis of the Concepts of Pollution and Taboo*. London: Routledge.

Douglas, S. (1985). The Navy adopts the radio: 1899–1919. In M. Smith (Ed.), *Military Enterprise and Technological Change: Perspectives on the American Experience* (pp. 117–173). Cambridge, Massachusetts: The MIT Press.

Dyens, O. (2001). *Metal and Flesh: The Evolution of Man: Technology Takes Over* (E. Bibbee and O. Dyens, Trans.). Cambridge, Massachusetts: The MIT Press.

Dyer, R. (1987). *Heavenly Bodies: Film Stars and Society*. London: Macmillan.

Dyer, R. (1998). *Stars* (New ed.). London: British Film Institute.

Early, J. E. (1996). Technology, modernity and the little man: Crippen's capture by wireless. *Victorian Studies, 39*(3), 309–339.

Easlea, B. (1983). *Fathering the Unthinkable: Masculinity, Scientists and the Nuclear Arms Race*. London: Pluto.

Edwards, P. (1987), A history of computers and weapons systems. In D. Bellin and G. Chapman (Eds), *Computers in Battle - Will they Work?* (pp. 45–60). Boston: Harcourt Brace Jovanovich.

Edwards, P. (1995). Cyberpunks in cyberspace: The politics of subjectivity in the computer age. In S. Star (Ed.), *The Cultures of Computing* (pp. 69–84). Oxford: Blackwell/Sociological Review.

Edwards, P. (1996). *The Closed World: Computers and the Politics of Discourse in Cold War America*. Cambridge, Massachusetts: The MIT Press.

Elliott, G. (2000). Reach for the sky. *The Australian*, May 13–14, 21, 28.

Ellis, A. (1998). Jasper F. Cropsey 1823–1900. In E. Johns, A. Sayers and E. Kornhauser (Eds), *New Worlds from Old: 19th Century Australian and American Landscape* (p. 172). Canberra: National Gallery of Australia/Hartford, Connecticut: Wadsworth Atheneum.

Essig, M. (2003). *Edison and the Electric Chair*. Camberwell: Penguin Australia.

Fanon, F. (1967). *The Wretched of the Earth* (C. Farrington, Trans.). Harmondsworth: Penguin.

Fisher, D. and M. Fisher (1996). *Tube: The Invention of Television*. Washington, DC: Centrepoint.

Fisher, J. (1997). The postmodern paradiso: Dante, cyberpunk, and the technosophy of cyberspace. In D. Porter (Ed.), *Internet Culture* (pp. 111–128). New York: Routledge.

Fishlock, T. (2004). *Conquerors of Time: Exploration and Invention in the Age of Daring*. London: John Murray.

Fist, S. (2001). Licensed to lose money. *The Australian*, May 15, 45.

Flew, A. (1983). Event. *A Dictionary of Philosophy* (2nd rev. ed.) (p. 115). London: Pan.

Flichy, P. (1995). *The Dynamics of Modern Communication: The Shaping and Impact of New Communication Technologies* (L. Libbrecht, Trans.). London: Sage.

Flint, R. W. (Ed.). (1972). *Marinetti: Selected Writings*. London: Secker & Warburg.

Foster, J. (2003). Capturing and losing the 'Lie of the land': Railway photography and colonialism nationalism in early twentieth-century South Africa. In J. Schwartz and J. Ryan (Eds), *Picturing Place: Photography and the Geographical Imagination* (pp. 141–161). London: I.B. Tauris.

Foucault, M. (1970). *The Order of Things: An Archaeology of the Human Sciences*. London: Tavistock.

Foucault, M. (1973). *The Birth of the Clinic: An Archaeology of Medical Perception*. (A. Sheridan, Trans.). London: Tavistock.

Foucault, M. (1977). *Discipline and Punish: The Birth of the Prison*. (A. Sheridan, Trans.). London: Penguin.

Foucault, M. (1980). The eye of power. In C. Gordon (Ed.), *Power/Knowledge: Selected Interviews and Other Writings 1972–1977* (pp. 146–165). Brighton: Harvester.

Foucault, M. (1988). *Technologies of the Self: A Seminar*. L. H. Martin, H. Gutman and P. H. Hutton (Eds). London: Tavistock.

Friedberg, A. (1993). *Window Shopping: Cinema and the Postmodern*. Berkeley: University of California Press.

Fuller, J. F. C. (1946/1998). *Armament and History: The Influence of Armament on History from the Dawn of Classical Warfare to the End of the Second World War*. New York: Da Capo.

Fuller, J. F. C. (1961/1992). *The Conduct of War 1789–1961: A Study of the Impact of the French, Industrial, and Russian Revolutions on War and its Conduct*. New York: Da Capo.

Fuller, J. F. C. (1943). *Machine Warfare: An Inquiry into the Influence of Mechanics on the Art of War*. Washington: Infantry Journal.

Furtwangler, A. (1993). The American sublime. *Acts of Discovery: Visions of America in the Lewis and Clark Journals* (pp. 23–51). Urbana: University of Illinois Press.

Galloway, J. (1972). *The Politics and Technology of Satellite Communications*. Lexington, Massachusetts: D.C. Heath.

Galvin, M. (1994). Vectory in the Gulf: Technology, communications and war. In L. Green and R. Guinery (Eds), *Framing Technology: Society, Choice and Change* (pp. 176–190). Sydney: Allen & Unwin.

Garratt, G. E. M. (1958). Telegraphy. In C. Singer *et al.* (Eds), *A History of Technology, volume IV: The Industrial Revolution c 1750 to c 1850* (pp. 644–662). Oxford: Clarendon Press.

Giblett, R. (1985). Watching TV, watching yourself: The viewer and the gaze. *Australian Journal of Cultural Studies*, 3(1), 120–127.

Giblett, R. (1996). *Postmodern Wetlands: Culture, History, Ecology*. Edinburgh: Edinburgh University Press.

Giblett, R. (1997a). Going green. *Continuum*, 11(2), 128–139.

Giblett, R. (1997b). Is the public sphere to the ecosphere as culture is to nature? (as male is to female?). *Continuum*, 11(3), 74–84.

Giblett, R. (2004). *Living with the Earth: Mastery to Mutuality*. Cambridge: Salt.

Giblett, R. (forthcoming). *The Body of the Earth*.

Gibson, W. (1984). *Neuromancer*. New York: Ace.

Gleick, J. (2002). *What just Happened: A Chronicle from the Information Frontier*. New York: Pantheon.

Gray, C. (1989). The cyborg soldier: The US military and the post-modern warrior. In L. Levidow and K. Robins (Eds), *Cyborg Worlds: The Military Information Society* (pp. 43–71). London: Free Association.

Grossman, K. (2001). Disgrace into space. *The Ecologist*, 31(2), 34–38.

Gunning, T. (1986). The cinema of attraction. *Wide Angle*, 8(3–4), 63–70.

Gunning, T. (1989). An aesthetic of astonishment: Early film and the (In)credulous spectator. *Art and Text*, 34, 31–45.

Hafner, K. and M. Lyon (1996). *Where Wizards Stay up Late: The Origins of the Internet*. New York: Simon & Schuster.

Hamelink, C. (1990). Information Imbalance: Core and Periphery. In J. Downing, A. Mohammadi and A. Sreberny-Mohammadi (Eds), *Questioning the Media: A Critical Introduction* (pp. 217–228). London: Sage.

Hammond, N. and H. Scullard (1970). (Eds). *The Oxford Classical Dictionary* (2nd ed.). Oxford: Clarendon Press.

Hanson, D. (1982). *The New Alchemists: Silicon Valley and the Microelectronics Revolution*. Boston: Little, Brown.

Haraway, D. (1985). A manifesto for cyborgs: Science, technology, and Socialist Feminism in the 1980s. *Socialist Review, 80*, 65–107.

Haraway, D. (1995). Cyborgs and symbionts: Living together in the new world order. In C. Gray (Ed.), *The Cyborg Handbook* (pp. xi–xx). London: Routledge.

Hauben, M. and R. Hauben (1997). *Netizens: On the History and Impact of Usenet and the Internet*. Los Alamitos, California: IEEE Computer Society Press.

Healy, D. (1997). Cyberspace and place: The Internet as middle landscape on the electronic frontier. In D. Porter (Ed.), *Internet Culture* (pp. 55–68). New York: Routledge.

Helmont, J. B. van (1650). *A Ternary of Paradoxes: Magnetick Cures of Wounds*. London: William Lee.

Herman, E. S. and R. W. McChesney (1997). *The Global Media: The New Missionaries of Corporate Capitalism*. London: Cassell.

Hindle, M. (1992). Introduction. To Mary Shelley, *Frankenstein, or the Modern Prometheus* (pp. vii–xliii). London: Penguin.

Hughes, R. (1980). *The Shock of the New: Art and the Century of Change*. London: British Broadcasting Corporation.

Hughes, T. (1983). *Networks of Power: Electrification in Western Society, 1880–1930*. Baltimore: Johns Hopkins University Press.

Hughes, T. (1998). *Rescuing Prometheus*. New York: Random House.

Huizinga, J. (1924). *The Waning of the Middle Ages: A Study of the Forms of Life, Thought, and Art in France and the Netherlands in the Fourteenth and Fifteenth Centuries* (F. Hopman, Trans.). Harmondsworth: Penguin.

Hutchinson, W. and M. Warren. (2001). Principles of information warfare. *Journal of Information Warfare, 1*(1), 1–6.

Huysmans, J.-K. (1959). *Against Nature* (R. Baldick, Trans.). London: Penguin.

Inglis, F. (1995). *Raymond Williams*. London: Routledge.

Irigaray, L. (1999). *The Forgetting of Air in Martin Heidegger* (M. Mader, Trans.). Austin: University of Texas Press.

Israel, P. (1992). *From Machine Shop to Industrial Laboratory: Telegraphy and the Changing Context of American Invention, 1830–1920*. Baltimore: The Johns Hopkins University Press.

Jagtenberg, T. and D. McKie (1997). *Eco-Impacts and the Greening of Postmodernity: New Maps for Communication Studies, Cultural Studies and Sociology*. London: Sage.

Jarry, A. (1902/1999). *The Supermale* (R. Gladstone and B. Wright, Trans.). Cambridge, Massachusetts: Exact Change.

Jarry, A. (1911/1996). *Exploits and Opinions of Doctor Faustroll, Pataphysician: A Neo-Scientific Novel* (S. Watson-Taylor, Trans.). Cambridge, Massachusetts: Exact Change.

Jassem, H. (1998). Regulating the electromagnetic spectrum: What happens when we're out of scarcity? *Paper Presented to the International Communication Association Annual Conference, Jerusalem, Israel*.

Jefferies, R. (2001). *At Home on the Earth: A New Selection of the Later Writings.* J. Hooker (Ed.). Totnes: Green Books.

Jenkins, R. (1987). Science, technology, and the evolution of photography, 1790–1925. In E. Ostroff (Ed.), *Pioneers of Photography: Their Achievements in Science and Technology.* Springfield: Society for Imaging Science & Technology.

Johnson, L. (1988). *The Unseen Voice: A Cultural Study of Early Australian Radio.* London: Routledge.

Jünger, E. (1920/2004). *Storm of Steel* (M. Hofmann, Trans.). London: Penguin.

Kaplan, A. (2000). The spectacle of war in Crane's revision of history. In D. Pizer (Ed.), S. Crane, *The Red Badge of Courage* (3rd ed., pp. 290–294). New York: Norton.

Kasson, J. (1976/1999). *Civilizing the Machine: Technology and Republican Values in America, 1776–1900.* New York: Hill & Wang.

Kattelle, A. (2000). *Home Movies: A History of the American Industry, 1897–1979.* Nashua, NH: Transition.

Kern, S. (1983). *The Culture of Time and Space, 1880–1918.* Cambridge, Massachusetts: Harvard University Press.

Kipling, R. (1904). 'Wireless'. *Traffics and Discoveries* (pp. 213–239). London: Macmillan.

Kirby, L. (1997). *Parallel Tracks: The Railroad and Silent Cinema.* Durham: Duke University Press.

Kirk, G., J. Raven and M. Schofield (1983). *The Presocratic Philosophers: A Critical History with a Selection of Texts* (2nd ed.). Cambridge: Cambridge University Press.

Kress, G. (1988). *Communication and Culture: An Introduction.* Kensington: University of New South Wales Press.

LaMay, C. and E. Dennis (Eds) (1991). *Media and the Environment.* Washington: Island Press.

Landa, M. de (1991). *War in the Age of Intelligent Machines.* New York: Zone.

Landes, D. (1969). *The Unbound Prometheus: Technological Change and Industrial Development in Western Europe from 1750 to the Present.* Cambridge: Cambridge University Press.

Laphlanche, J. and J-B. Pontallis (1973). *The Language of Psychoanalysis.* (D. Nicholson-Smith, Trans.). London: Hogarth.

Lawson, H. (1976). The roaring days. In B. Kiernan (Ed.), *Portable Australian Authors: Henry Lawson* (pp. 77–78). St Lucia: University of Queensland Press.

Lax, S. (1997). *Beyond the Horizon: Communications Technologies: Past, Present and Future.* Luton: University of Luton Press.

Leach, W. (1993). *Land of Desire: Merchants, Power and the Rise of a New American Culture.* New York: Random House.

Levidow, L. and K. Robins (Eds). (1989a). Introduction. *Cyborg Worlds: The Military Information Society* (pp. 7–11). London: Free Association.

Levidow, L. and K. Robins (Eds) (1989b). Towards a military information society? *Cyborg Worlds: The Military Information Society* (pp. 159–177). London: Free Association.

Levin, H. (1971). *The Invisible Resource: Use and Regulation of the Radio Spectrum.* Baltimore: Johns Hopkins Press for Resources for the Future.

Lévy, P. (2001). *Cyberculture* (R. Bononno, Trans.). Minneapolis: University of Minnesota Press.

Lewis, J. (1996). *Mining the Sky: Untold Riches from the Asteroids, Comets, and Planets.* Reading, Massachusetts: Addison-Wesley.

Licklider, J. C. R. (1960). Man-computer symbiosis. http://memx.org.licklider. html

Licklider, J. C. R. (1965). *Libraries of the Future.* Cambridge, Massachusetts: The MIT Press.

Licklider, J. C. R. and R. W. Taylor (1968). The computer as a communication device. http://memx.org.licklider.html

Lines, W. J. (1998). *False Economy: Australia in the Twentieth Century.* South Fremantle: Fremantle Arts Centre Press.

L'Isle-Adam, V. de (1963). Celestial publicity. *Cruel Tales* (pp. 41–45) (R. Baldick, Trans.). London: Oxford University Press.

L'Isle-Adam, V. de (1981). *The Eve of the Future Eden* (M. Rose, Trans.). Lawrence, Kansas: Coronado.

Littauer, R. and Uphoff, N. (1972). (Eds). *The Air War in Indochina* (rev. ed.). Boston: Beacon Press.

Livingston, K. (1996). *The Wired Nation Continent: The Communication Revolution and Federating Australia.* Melbourne: Oxford University Press.

Lloyd, G. and N. Sivin (2002). *The Way and the Word: Science and Medicine in Early China and Greece.* New Haven: Yale University Press.

Locke, J. (1690/1947). *An Essay Concerning Human Understanding* R. Wilburn (Ed.). London: Dent.

Lodge, O. (1883). The ether and its functions. *Nature, XXVII*, 328–330.

Lyotard, J.-F. (1989). The sublime and the avant-garde. In A. Benjamin (Ed.), *The Lyotard Reader.* Oxford: Basil Blackwell.

MacKechnie Jarvis, C. (1958). Telegraphs. In C. Singer *et al.* (Eds), *A History of Technology, volume V: The Late Nineteenth Century c 1850 to c 1900* (pp. 218–224). Oxford: Clarendon Press.

McLuhan, M. (1964). *Understanding Media: The Extensions of Man.* London: Routledge & Kegan Paul.

McLuhan, M. (1967). *The Mechanical Bride: Folklore of Industrial Man.* Boston: Beacon.

Man Chan, J. (1997). National responses and accessibility to STAR TV in Asia. In O. Boyd-Barrett, A. Sreberny-Mohammadi, D. Winseck and J. McKenna (Eds), *Media in Global Context: A Reader* (pp. 94–106). London: Arnold.

Marshall, A. (1999). Gaining a share of the final frontier. In B. Martin (Ed.), *Technology and Public Participation.* Wollongong: University of Wollongong Press.

Martin, A. (1992). *Railroads Triumphant: The Growth, Rejection, and Rebirth of a Vital American Force.* New York: Oxford University Press.

Marvin, C. (1987). Information and history. In J. Slack and F. Fejes (Eds), *The Ideology of the Information Age* (pp. 49–62). Norwood, NJ: Ablex.

Marvin, C. (1988). *When Old Technologies were New: Thinking About Electric Communication in the Late Nineteenth Century.* New York: Oxford University Press.

Marx, K. (1986). *Economic Manuscripts of 1857–1858 (Grundrisse), Collected Works, volume 28.* London: Lawrence & Wishart.

Marx, K. (1987). *Economic Manuscripts of 1857–1858 (Grundrisse), Collected Works, volume 29.* London: Lawrence & Wishart.

Marx, L. (1964). *The Machine in the Garden: Technology and the Pastoral Ideal in America.* New York: Oxford University Press.

Marx, L. (1988a). Literature, technology, and covert culture. *The Pilot and the Passenger: Essays on Literature, Technology, and Culture in the United States* (pp. 133–135). New York: Oxford University Press.

Marx, L. (1988b). The railroad-in-the-landscape: An iconological reading of a theme in American art. In L. Marx and S. Danly (Eds), *The Railroad in American Art: Representations of Technological Change* (pp. 183–209). Cambridge, Massachusetts: The MIT Press.

Melville, H. (1950). The paradise of bachelors and the tartarus of maids. In R. Chase (Ed.), *Selected Tales and Poems* (pp. 206–229). New York: Holt, Rinehart & Winston.

Metz, C. (1982). *Psychoanalysis and Cinema: The Imaginary Signifier* (C. Britton, A. Williams, B. Brewster and A. Guzzetti, Trans.). London: Macmillan.

Metzger, T. (1996). *Blood and Volts: Edison, Tesla, and the Electric Chair*. Brooklyn: Autonomedia.

Michaelis, A. R. (1978). Space technology. In T. I. Williams (Ed.), *A History of Technology, Volume VII: The Twentieth Century c.1900 to c.1950* II (pp. 857–870). Oxford: Clarendon Press.

Michelsen, A. (1984). On the Eve of the future: The reasonable facsimile and the philosophical toy. *October, 29*, 3–21.

Miller, P. (1966). Technological America. *The Life of the Mind in America: From the Revolution to the Civil War* (pp. 269–313). London: Victor Gollancz.

Miller, T. (1997). Radio. In S. Cunningham and G. Turner (Eds), *The Media in Australia* (2nd ed., pp. 47–69). Sydney: Allen & Unwin.

Morus, I. (1996). The electric Ariel: Telegraphy and commercial culture in early Victorian England. *Victorian Studies, 39*(3), 339–378.

Morus, I. (1998). To annihilate time and space: The invention of the telegraph. *Frankenstein's Children: Electricity, Exhibition, and Experiment in Early-Nineteenth-Century London* (pp. 194–230). Princeton, New Jersey: Princeton University Press.

Mosco, V. (2004). *The Digital Sublime: Myth, Power and Cyberspace*. Cambridge, MA: MIT Press.

Mowlana, H. (1997). *Global Information and World Communication: New Frontiers in International Relations* (2nd ed). London: Sage.

Moyal, A. (1984). *Clear Across Australia: A History of Telecommunications*. Melbourne: Nelson.

Neale, S. (1985). *Cinema and Technology: Image, Sound, Colour*. London: Macmillan.

Nicholl, C. (1997). *Journeys*. London: Dent.

Nicholls, B. (1996). The work of culture in the age of cybernetic systems. In T. Druckrey (Ed.), *Electronic Culture: Technology and Visual Representation* (pp. 121–143). New York: Aperture.

Noakes, R. (1999). Telegraphy is an occult art: Cromwell Fleetwood and the diffusion of electricity to the other world. *British Journal for the History of Science, 32*(115), 421–459.

Noam, E. (1998). Spectrum auctions: Yesterday's heresy, today's orthodoxy, tomorrow's anachronism. Taking the next step to open spectrum access. *Journal of Law and Economics, XLI*, 765–790.

Noble, D. (1984). *Forces of Production: A Social History of Industrial Automation*. New York: Knopf.

Noble, D. (1999). *The Religion of Technology: The Divinity of Man and the Spirit of Invention*. New York: Penguin.

Novak, B. (1980). *Nature and Culture: American Landscape and Painting 1825–1875*. London: Thames & Hudson.

Nye, D. (1994). *American Technological Sublime*. Cambridge, Massachusetts: The MIT Press.

Oberg, J. (nd). *Space power theory*. http://www.spacecom.af.mil/usspace/

Oberg, J. (2001). The heavens at war. *New Scientist*, 2 June, 26–29.

O'Dell, A. and P. Richards (1971). *Railways and Geography* (2nd ed.). London: Hutchinson.

Oettermann, S. (1997). *The Panorama: History of a Mass Medium* (D. Schneider, Trans.). New York: Zone.

Okolie, C. (1989). *International Law of Satellite Remote Sensing and Outer Space*. Dubuque: Kendall/Hunt.

Olsen, L. (1991). The shadow of spirit in William Gibson's matrix trilogy. *Extrapolations, 32*(3), 278–289.

Pagel, W. (1982). *Joan Baptista Van Helmont: Reformer of Science and Medicine*. Cambridge: Cambridge University Press.

Pickett, C. (1998). *Cars and Culture: Our Driving Passions*. Sydney: Powerhouse Publishing & Harper/Collins.

Plato (1969). Phaedo. *The Last Days of Socrates* (H. Tredennick, Trans.). London: Penguin.

Pocock, R. (1988). *The Early British Radio Industry*. Manchester: Manchester University Press.

Redhead, S. (2004). *Paul Virilio: Theorist for an Accelerated Culture*. Edinburgh: Edinburgh University Press.

Remarque, E. (1929/1994). *All Quiet on the Western Front* (B. Murdoch, Trans.). London: Vintage.

Riordan, M. and L. Hoddeson (1997). *Crystal Fire: The Birth of the Information Age*. New York: W. W. Norton.

Robins, K. (1994). The haunted screen. In G. Bender and T. Druckrey (Eds), *Culture on the Brink: Ideologies of Technology* (pp. 305–315). Seattle: Bay Press.

Robins, K. (1996a). Cyberspace and the world we live in. In J. Dovey (Ed.), *Fractal Dreams: New Media in Social Context* (pp. 1–30). London: Lawrence & Wishart.

Robins, K. (1996b). *Into the Image: Culture and Politics in the Field of Vision*. London: Routledge.

Robins, K. (1996c). The virtual unconscious in postphotography. In T. Druckrey (Ed.), *Electronic Culture: Technology and Visual Representation* (pp. 154–163). New York: Aperture.

Robins, K. and L. Levidow (1991). The eye of the storm. *Screen, 32*(3), 324–328.

Robins, K. and L. Levidow (1995a). Socializing the cyborg self: The Gulf War and beyond. In C. Gray (Ed.), *The Cyborg Handbook* (pp. 119–125). New York: Routledge.

Robins, K. and L. Levidow (1995b). Soldier, cyborg, citizen. In J.Brock and I. Boal (Eds). *Resisting the Virtual Life: The Culture and Politics of Information* (pp. 105–113). San Francisco: City lights.

Robins, K. and F. Webster (1999). *Times of the Technoculture: From the Information Society to the Virtual Life*. London: Routledge.

Rose, H. (1928). *A Handbook of Greek Mythology*. London: Methuen.

Roszak, T. (1988). *The Cult of Information: The Folklore of Computers and the True Art of Thinking*. London: Paladin.

Ryan, J. (1997). *Picturing Empire: Photography and the Visualization of the British Empire*. London: Reaktion Books.

Sachs, W. (1992). *For love of the Automobile: Looking Back in the History of our Desires* (D. Reneau, Trans.). Berkeley: University of California Press.

Saussure, F. de (1974). *Course in General Linguistics* C. Bally and A. Sechehaye (Eds), (W. Baskin, Trans.). London: Fontana.

Schaffer, S. (1996). Babbage's dancer and the impresarios of mechanism. In J. Uglow and F. Spufford (Eds), *Cultural Babbage: Technology, Time and Invention* (pp. 53–80). London: Faber & Faber.

Schivelbusch, W. (1986). *The Railway Journey: The Industrialization of Time and Space in the 19th Century*. Leamington Spa: Berg.

Schivelbusch, W. (1988). *Disenchanted Night: The Industrialization of Light in the Nineteenth Century*. (A. Davies, Trans.). Oxford: Berg.

Schwartz, J. and J. Ryan (Eds) (2003). Introduction: Photography and the geographical imagination. *Picturing Place: Photography and the Geographical Imagination* (pp. 1–18). London: I. B. Tauris.

Sconce, J. (2000). *Haunted Media: Electronic Presence from Telegraphy to Television*. Durham: Duke University Press.

Seddon, G. (1997). *Landprints: Reflections on Place and Landscape*. Melbourne: Cambridge University Press.

Sekula, A. (1984). The instrumental image: Steichen at war. *Photography Against the Grain: Essays and Photo Works 1973–1983* (pp. 33–51). Halifax, Nova Scotia: Press of the Nova Scotia College of Art & Design.

Serres, M. (1982a). *Hermes: Literature, Science, Philosophy* J. Harari and D. Bell (Eds). Baltimore: The Johns Hopkins University Press.

Serres, M. (1982b). *The Parasite* (L. Schehr, Trans.). Baltimore: The Johns Hopkins University Press.

Serres, M. and Michelet (1995a). *Genesis* (G. James and J. Nielson, Trans.). Ann Arbor: University of Michigan Press.

Serres, M. (1995b). *The Natural Contract*. (E. MacArthur and W. Paulson, Trans.). Ann Arbor: University of Michigan Press.

Serres, M. with B. Latour (1995). *Conversations on Science, Culture, and Time* (R. Lapidus, Trans.). Ann Arbor: University of Michigan Press.

Sheehy, C. (2001). Space warriors defend information assets. *Signal*. http://www.us.net/signal/CurrentIssue/April01/space-april.html

Shelley, M. (1818/1992). *Frankenstein, or the Modern Prometheus*. London: Penguin.

Shurkin, J. (1984). *Engines of the Mind: A History of the Computer*. New York: W. W. Norton.

Slotkin, R (1973). *Regeneration Through Violence: The Mythology of the American Frontier, 1600–1860*. Hanover: Wesleyan University Press.

Slotkin, R. (1992). *Gunfighter Nation: The Myth of the Frontier in Twentieth-Century America*. New York: HarperCollins.

Slotkin, R. (1994). *The Fatal Environment: The Myth of the Frontier in the Age of Industrialization 1800–1890*. New York: HarperCollins.

Smith, H. (1950). *Virgin Land: The American West as Symbol and Myth*. Cambridge, Massachusetts: Harvard University Press.

Smith, M. (Ed.) (1985). Introduction. *Military Enterprise and Technological Change: Perspectives on the American Eexperience* (pp. 1–37). Cambridge, Massachusetts: The MIT Press.

Smith-Rose, R. (1948). *James Clerk Maxwell, F. R. S.: 1831–1879*. London: Longmans, Green.

Smythe, D. (1981). The electronic information tiger, or the political economy of the radio spectrum and the Third World interest. *Dependency Road: Communications, Capitalism, Consciousness and Canada* (pp. 300–318). Norwood, NJ: Ablex.

Sofia, Z. (1984). Exterminating fetuses: Abortion, disarmament, and sexosemiotics of extraterrestrialism. *Diacritics, 14*(2), 47–59.

Sofia, Z. (1990). Spacing out in the mother shop. In T-A. White, A. Gibbs, W. Jenkins and N. King (Eds), *No Substitute: Prose Poems Images* (pp. 133–143). Fremantle: Fremantle Arts Centre Press.

Sofia, Z. (1992). Hegemonic irrationalities and psychoanalytic cultural critique. *Cultural Studies, 6*(3), 376–394.

Sofia, Z. (1996). Contested zones: Futurity and technological art. *Leonardo, 29*(1), 59–66.

Sofoulis, Z. (1983). *Alien Pre-Oedipus*: Penis-breast, cannibaleyes. *Presented as Part of the Qualifying Essay, History of Consciousness, University of California, Santa Cruz.*

Sofoulis, Z. (1990). Semiotics of technology. *Lecture for Social Semiotics, Murdoch University.*

Sontag, S. (1977). *On Photography*. New York: Farrar, Straus & Giroux.

Spretnak, C. (1997). *The Resurgence of the Real: Body, Nature and Place in a Hypermodern World*. Reading, Massachusetts: Addison-Wesley.

Standage, T. (1998). *The Victorian Internet: The Remarkable Story of the Telegraph and the Nineteenth Century's Online Pioneers*. London: Weidenfeld & Nicolson.

Stone, A. (1996). *The War of Desire and Technology at the Close of the Mechanical Age*. Cambridge, Massachusetts: The MIT Press.

Stratton, J. (1997). Cyberspace and the globalization of culture. In D. Porter (Ed.), *Internet Culture* (pp. 253–275). London: Routledge.

Streeter, T. (2000). Notes towards a political history of the Internet 1950–1983. *Media International Australia, 95*, 131–146.

Suzuki, D. (1998). *Earth Time*. St Leonards: Allen & Unwin.

Swade, D. (1996). 'It will not slice a pineapple': Babbage, miracles and machines. In J. Uglow and F. Spufford (Eds). *Cultural Babbage: Technology, Time and Invention* (pp. 34–51). London: Faber & Faber.

Takeguchi, E. and W. Wooley (1992). Spectrum management. In A. Campen (Ed.), *The First Information War: The Story of Communications, Computers and Intelligence in the Persian Gulf War* (pp. 155–160). Fairfax: AFCEA International Press.

Taylor, N. (1973). The awful sublimity of the Victorian city: Its aesthetic and architectural origins. In H. Dyos and M. Wolff (Eds), *The Victorian City: Images and Realities, volume 2* (pp. 431–447). London: Routledge & Kegan Paul.

Taylor, R. (1975). *Electricity*. Harmondsworth: Penguin.

Theweleit, K. (1987). *Male Fantasies, volume I: Women, Floods, Bodies, History* (S. Conway, Trans.). Cambridge: Polity Press.

Theweleit, K. (1989). *Male Fantasies, volume II, Male Bodies: Psychoanalyzing the White Terror* (C. Turner and E. Carter, Trans.). Cambridge: Polity.

Thompson, E. (1968). *The Making of the English Working Class* (rev. ed.). London: Penguin.

Thompson, E. (1993). Time, work-discipline and industrial capitalism. *Customs in Common* (pp. 352–403). London: Penguin.
Thompson, J. (1995). *The Media and Modernity: A Social Theory of the Media.* Cambridge: Polity.
Thompson, R. (1947). *Wiring a Continent: The History of the Telegraph Industry in the United States 1832–1866.* Princeton: Princeton University Press.
Thomson, A. (1999). *The Singing Line.* London: Chatto & Windus.
Thoreau, H. D. (1854/1997). *Walden.* Boston: Beacon.
Tillyard, E. M. W. (1943). *The Elizabethan World Picture.* Harmondsworth: Penguin.
Toffler, A. H. (1995). *War and Anti-War.* New York: Time Warner.
Toma, J. (1992). Desert Storm communications. In A. Campen (Ed.), *The First Information War: The Story of Communications, Computers and Intelligence in the Persian Gulf War* (pp. 1–5). Fairfax: AFCEA International Press.
Tucker, D. (1978). Electrical Communication. In T. Williams (Ed.), *A History of Technology, volume VII: The Twentieth Century c.1900 to c.1950* II (pp. 1220–1267). Oxford: Clarendon Press.
Turner, F. (1893/1961). The significance of the frontier in American history. *Frontier and Section: Selected Essays of Frederick Jackson Turner* (pp. 37–62). Englewood Cliffs: Prentice-Hall.
Uglow, J. (1996). Introduction: Possibility. In J. Uglow and F. Spufford (Eds), *Cultural Babbage: Technology, Time and Invention* (pp. 1–23). London: Faber & Faber.
United States Space Command http://www.spacecom.af.mil/usspace/
Vernadsky, W. I. (1945). The Biosphere and the Noösphere. *American Scientist, 33*(1), 1–12.
Verne, J. (1889/1999). In the twenty-ninth century: The day of an American journalist in the year 2889. *The Eternal Adam and Other Stories* (pp. 191–206). London: Phoenix.
Vertov, D. (1984). *Kino-Eye: Writings.* A. Michelsen (Ed.), (K. O'Brien, Trans.). Berkeley: University of California Press.
Virilio, P. (1986a). The overexposed city. *Zone, 1/2,* 15–31.
Virilio, P. (1986b). *Speed and Politics: An Essay on Dromology* (M. Polizzotti, Trans.). New York: Semiotext(e).
Virilio, P. (1988). The third window: An interview. In C. Schneider and B. Wallis (Eds), *Global Television* (pp. 187–197). New York: Wedge.
Virilio, P. (1989a). The last vehicle. In D. Kamper and C. Wulf (Eds), (D. Antal, Trans.), *Looking Back on the End of the World* (pp. 106–119). New York: Semiotext (e).
Virilio, P. (1989b). *War and Cinema: The Logistics of Perception* (P. Camiller, Trans.). London: Verso.
Virilio, P. (1990). *Popular Defense and Ecological Struggles* (M. Polizzotti, Trans.). New York: Semiotext (e).
Virilio, P. (1991a). *The Aesthetics of Disappearance* (P. Beitchman, Trans.). New York: Semiotext (e).
Virilio, P. (1991b). *The Lost Dimension* (D. Moshenberg, Trans.). New York: Semiotext (e).
Virilio, P. (1993). The third interval: A critical transition. In V. A. Conley (Ed.), *Rethinking Technologies* (pp. 3–12). Minneapolis: University of Minnesota Press.

Virilio, P. (1994a). *Bunker Archaeology* (G. Collins, Trans.). New York: Princeton Architectural Press.

Virilio, P. (1994b). *The Vision Machine* (J. Rose, Trans.). Bloomington: Indiana University Press.

Virilio, P. (1995). *The Art of the Motor* (J. Rose, Trans.). Minneapolis: University of Minnesota Press.

Virilio, P. (1997a). Cyberwar, God and television: An interview. In Arthur & M. Kroker (Eds). *Digital Delirium* (pp. 41–48). New York: St Martin's.

Virilio, P. (1997b). *Open sky* (J. Rose, Trans.). London: Verso.

Virilio, P. (1998). Desert screen. In J. Der Derian (Ed.), *The Virilio Reader* (pp. 166–182). Oxford: Blackwell.

Virilio, P. (1999). *Politics of the Very Worst: An Interview by Philippe Petit*. Sylvère Lotringer (Ed.), (M. Cavaliere, Trans.). New York: Semiotext (e).

Virilio, P. (2000a). *The Information Bomb* (C. Turner, Trans.). London: Verso.

Virilio, P. (2000b). *A Landscape of Events* (J. Rose, Trans.). Cambridge, Massachusetts: The MIT Press.

Virilio, P. (2000c). *Polar Inertia* (P. Camiller, Trans.). London: Sage.

Virilio, P. (2000d). *Strategy of Deception* (C. Turner, Trans.). London: Verso.

Virilio, P. (2000e). Twilight of the grounds. *The Desert* (pp. 102–118). London: Thames & Hudson.

Virilio, P. (2001). *Virilio live: Selected Interviews* J. Armitage (Ed.). London: Sage.

Virilio, P. (2002a). *Desert Screen: War at the Speed of Light* (M. Degener, Trans.). London: Continuum.

Virilio, P. (2002b). *Ground Zero* (C. Turner, Trans.). London: Verso.

Virilio, P. (2003a). *Art and Fear* (J. Rose, Trans.). London: Continuum.

Virilio, P. (2003b). *Unknown Quantity* (C. Turner, Trans.). London: Thames & Hudson.

Virilio, P. (2005a). *City of Panic* (J. Rose, Trans.). Oxford: Berg.

Virilio, P. (2005b). *Negative Horizon* (M. Degener, Trans.). London: Continuum.

Virilio, P. (2007). *The Original Accident* (J. Rose, Trans.). Cambridge: Polity.

Virilio, P. and S. Lotringer (1983). *Pure War* (M. Polizzotti, Trans.). New York: Semiotext(e).

Virilio, P. and S. Lotringer (2002). *Crepuscular Dawn* (M. Taormina, Trans.). New York: Semiotext(e).

Virilio, P. and S. Lotringer (2005). *The Accident of Art*. New York: Semiotext(e).

Voller, J. (1993). Neuromanticism: Cyberspace and the sublime. *Extrapolations*, *34*(1), 18–29.

Voloshinov, V. (1973). *Marxism and the Philosophy of Language* (L. Matejka and I. R. Titunik, Trans.). New York: Seminar Press.

Wark, McK. (1994). Third Nature. *Cultural Studies*, *8*(1), 115–132.

Weaver, W. (1949). Recent contributions to the mathematical theory of communication. In W. Weaver and C. Shannon, *The Mathematical Theory of Communication* (pp. 3–28). Urbana: University of Illinois Press.

Weizenbaum, J. (1976). *Computer Power and Human Reason: From Judgment to Calculation*. San Francisco: W.H. Freeman.

Wertheim, M. (1999). *The Pearly Gates of Cyberspace: A History of Space from Dante to the Internet*. Sydney: Doubleday.

Whitman, W. (1945). *The Portable Walt Whitman*. M. Van Doren (Ed.). New York: Penguin.

Whitney-Smith, E. (1996). War, information and history: Changing paradigms. In A. Campen, D. Dearth and R. Gooden (Eds), *Cyberwar: Security, Strategy and Conflict in the Information Age* (pp. 53–69). Fairfax, VA: AFCEA International Press.

Whittaker, E. (1989). *A History of the Theories of Aether and Electricity*. New York: Dover.

Wiener, N. (1954/1989). *The Human Use of Human Beings: Cybernetics and Society*. London: Free Association.

Williams, H. (1991). *Autogeddon*. London: Jonathan Cape.

Williams, R. (1958/2001). Culture is ordinary. In J. Higgins (Ed.), *The Raymond Williams Reader* (pp. 10–24). Oxford: Blackwell.

Williams, R. (1968). *Communications* (rev. ed.). Harmondsworth: Penguin.

Williams, R. (1973). *The Country and the City*. London: Chatto & Windus.

Williams, R. (1974). *Television: Technology and Cultural Form*. Hanover, New Hampshire: Wesleyan University Press.

Williams, R. (1976). Communication. *Keywords: A Vocabulary of Culture and Society* (pp. 62–63). Glasgow: Fontana.

Williams, R. (1980a). Advertising: The magic system. *Problems in Materialism and Culture: Selected Essays* (pp. 170–195). London: NLB.

Williams, R. (1980b). Means of communication as means of production. *Problems in Materialism and Culture: Selected Essays* (pp. 50–63). London: NLB.

Williams, R. (1981). Communication technologies and social institutions. In R. Williams (Ed.), *Contact: Human Communication and its History* (pp. 226–238). London: Thames & Hudson.

Williams, R. (1985). *Towards 2000*. Harmondsworth: Penguin.

Williams, R. (1989). *The Politics of Modernism: Against the New Conformists*. London: Verso.

Williams, R. H. (1982). *Dream Worlds: Mass Consumption in Late Nineteenth-Century France*. Berkeley: University of California Press.

Williams, R. H. (1990). *Notes on the Underground: An Essay on Technology, Society and the Imagination*. Cambridge, Massachusetts: The MIT Press.

Williams, R. H. (2000). 'All that is solid melts into air': Historians of technology in the information revolution. *Technology and Culture, 41*, 641–668.

Wilson, A. (1992). *The Culture of Nature: North American Landscape from Disney to the Exxon Valdez*. Cambridge, Massachusetts: Blackwell.

Wilton, A. and T. Barringer (2002). *American Sublime: Landscape Painting in the United States 1820–1880*. Princeton: Princeton University Press.

Winston, B. (1998). *Media Technology and Society: A History: From the Telegraph to the Internet*. London: Routledge.

Wolin, R. (1982). *Walter Benjamin: An Aesthetic of Redemption*. New York: Columbia University Press.

Wordsworth, W. (1906). Kendal and Windermere Railway. *Guide to Lakes: Fifth Edition (1835)* (pp. 146–166). Oxford: Oxford University Press.

Wordsworth, W. (1972). *The Prelude: A Parallel Text* J. Maxwell (Ed.). London: Penguin.

Young, P. and P. Jesser (1997). *The Media and the Military: From the Crimea to Desert Strike*. New York: St. Martin's Press.

Zuboff, S. (1988). *In the Age of the Smart Machine: The Future of Work and Power*. New York: Basic Books.

Index